A GOVERNANCE APPROACH TO URBAN WATER PUBLIC–PRIVATE PARTNERSHIPS

CASE STUDIES AND LESSONS FROM ASIA AND THE PACIFIC

MARCH 2022

ASIAN DEVELOPMENT BANK

Contents

Tables, Figures, and Boxes

Foreword

The Asian Water Development Outlook 2020: Advancing Water Security Across Asia and the Pacific reveals that 600 million people in the region's urban areas do not have access to adequate water or sanitation. According to Strategy 2030 of the Asian Development Bank (ADB), Asia and the Pacific needs $53 billion per year in water investment through 2030.

The private sector will need to provide about one-third of this amount. The region has attained remarkable growth over the past 10 years. However, equitable access to clean water and wastewater treatment to improve public health is lacking. Good water sector governance is essential to mobilize public and private finance, especially during the coronavirus disease (COVID-19) pandemic.

This publication aims to support Operational Priority 6 of ADB's Strategy 2030—to strengthen governance and institutional capacity—and the G20 Principles for Quality Infrastructure Investment (QII). It relates in particular to QII Principle 2, which seeks to increase the economic efficiency of infrastructure through robust project appraisal, including the assessment of life-cycle costs, fiscal sustainability, and affordability. It is also relevant to Principle 6, which focuses on governance of the project cycle from procurement to asset management. In addition, the publication reflects Sustainable Development Goal 6 on access to clean water and sanitation for all.

This report provides practical guidance and advice on the governance of public–private partnerships (PPPs) to support government decision-makers and private practitioners in the water and sanitation sectors. The report also provides a governance perspective based on the operational experience and knowledge of ADB in the water sector. It employs a broad definition of PPPs in the context of Asia and the Pacific's diverse social and economic conditions.

Seven PPP case studies in Asia are presented—covering water distribution and wastewater treatment. These case studies show that ADB developing member countries require a public policy that is committed to achieving water security that is inclusive and accessible to all. They emphasize the need for public institutions to have the capacity to prepare, monitor, and oversee projects. The case studies also demonstrate that transparent subsidies and effective fiscal management can ensure sustainable funding.

Improving governance is one of Asia and the Pacific's most critical challenges. Addressing this challenge can maximize private sector participation, build resilience and adaptive capacity, and improve access to water, especially for poor and vulnerable populations.

Hiranya Mukhopadhyay
Chief of the Governance Thematic Group
Sustainable Development and Climate Change Department
Asian Development Bank

Acknowledgments

The preparation of this report was led by Hanif Rahemtulla, principal public management specialist, Governance Thematic Group of the Asian Development Bank's Sustainable Development and Climate Change Department (SDTC-GOV). The co-lead writers were Anand Madhavan (director, CRISIL Risk and Infrastructure Solutions Limited) and David Bloomgarden (PPP governance specialist and ADB consultant, SDTC-GOV).

The authors thank the following colleagues for providing strategic guidance to ensure this report reflects the operational experience and knowledge of ADB in water sector PPPs: Jingmin Huang, director, Urban Development and Water Division of the Pacific Department (PAUW); Norio Saito, director, Urban Development and Water Division of the South Asia Department (SAUW); Ye Yong, country director, Pakistan Resident Mission; and Vijay Padmanabhan, director, Operations Planning and Coordination Division of the Strategy, Policy, and Partnerships Department.

The project team wishes to acknowledge the technical feedback and support for the case studies provided by Pratish Halady, senior advisor to the managing director general, Office of the President; Ron H. Slangen, unit head, portfolio management, Urban Development and Water Division of the Central and West Asia Department (CWUW); Allison Woodruff, principal water security specialist, Water Sector Group, Sector Advisory Services Cluster of the Sustainable Development and Climate Change Department; Joris G. Van Etten, senior urban development specialist, Urban Development and Water Division of the Southeast Asia Department (SEUW); Sanjay Divakar Joshi, principal urban development specialist, SAUW; Momoko Tada, senior urban development specialist, SAUW; Sanjay Grover, senior public-private partnership specialist, PPP-Thematic Group Secretariat; Jung Ho Kim, senior urban development specialist, CWUW; Cesar Llorens, senior urban development specialist, CWUW; Vivian Castro-Wooldridge, senior urban development specialist, PAUW; Vivek Rao, principal financial sector specialist, Public Management, Financial Sector, and Trade Division of the Southeast Asia Department; and Joao Pedro Farinha Fernandes, principal financial sector economist, Public Management, Financial Sector, and Trade Division of the Central and West Asia Department.

A draft of this publication was presented in a July 2021 workshop on Strengthening Fiscal Governance and Sustainability for PPPs in the Urban Water Sector, organized by ADB and BAPPENAS (Ministry of National Development Planning, Indonesia). The authors also wish to acknowledge the Ministry of Foreign Affairs of India and the Ministry of Finance of Armenia for their review of the report.

Abbreviations

Acea	Electricity and Water Municipal Utility (Azienda Comunale Elettricità e Acque)
ADB	Asian Development Bank
AWDO	Asian Water Development Outlook
AWSC	Armenia Water Services Company
BOQ	bill of quantity
BOT	build–operate–transfer
CMC	Coimbatore Municipal Corporation
COVID-19	coronavirus disease
DBO	design–build–operate
DJB	Delhi Jal Board
GIS	geographic information system
IIGF	Indonesia Infrastructure Guarantee Fund
JWSRB	Jakarta Water Supply Regulatory Body
KPI	key performance indicator
KUIDFC	Karnataka Urban Infrastructure Development and Finance Corporation
LAC	Latin America and the Caribbean
LPCD	liters per capita per day
LWSC	Lori Water Services Company
NAWSC	Nor Akunq Water Services Company
NMCG	National Mission for Clean Ganga (India)
NRW	non-revenue water
O&M	operations and maintenance
PBCOC	performance-based construct and operate contract
PDAB	Indonesian provincially owned water company
PDAM	Perusahaan Daerah Air Minum (Indonesian regional water utility company)
PPI	private participation in infrastructure
PPP	public–private partnership
PRC	People's Republic of China
PSRC	Public Services Regulatory Commission
PT SMI	Indonesian infrastructure company (PT Sarana Multi Infrastruktur)
QII	Quality Infrastructure Investment
SCWE	State Committee for Water Economy, Armenia
SDGs	Sustainable Development Goals
STP	sewage treatment plant
SWSC	Shirak Water Services Company
TNIDB	Tamil Nadu Infrastructure Development Board
UFW	unaccounted for water
YWSC	Yerevan Water Services Company

About the Authors

Hanif Rahemtulla, Principal Public Management Specialist, Governance Thematic Group, ADB

Hanif Rahemtulla joined ADB in 2017, and has focused on leading and contributing to operational engagements in public investment management for better service delivery. Prior to joining ADB, Hanif was a Senior Operations Officer at the World Bank Group (2010–2017) and internationally assigned to the Philippines (2014–2017), where he led and contributed to economic sector work, technical assistance, and operations in public sector management. He has led and contributed to operations in India, Viet Nam, Indonesia, Tajikistan, and Mongolia. In this capacity, he supported country objectives such as addressing the rural infrastructure gap, and enhancing agricultural productivity and the agribusiness sector. He is a national of the United Kingdom. He obtained his doctoral degree from University College London (UCL) and was a postdoctoral fellow at Canada's McGill University in 2009–2010.

Anand Madhavan, Director, Infrastructure & Public Finance, CRISIL Risk and Infrastructure Solutions Limited, India; Consultant to ADB

Anand brings 20 years of experience in infrastructure and development consulting. He has worked with governments, multilateral institutions, investors, and private developers in the areas of infrastructure policy, reforms, project development, and transaction advisory across India and other emerging economies in Asia and Africa. Anand leads consulting engagements in public–private partnerships, infrastructure finance, and urban finance across India, Asia, and Africa. He is a contributor and coauthor for the ADB-led publication *Good Practices in Urban Water Management: Decoding Good Practices for a Successful Future*, which details best practices in urban water management in eight Asian cities. He has led the preparation of other publications and tool kits, including a reference tool on infrastructure project preparation, on behalf of the G20 Global Infrastructure Hub, and a guidance framework for the issuance of municipal bonds on behalf of the Government of India.

 **David Bloomgarden**, Public–Private Partnership and Governance Expert and Consultant to ADB

David Bloomgarden has over 30 years of global experience in policy, management, and project design and implementation. At the Inter-American Development Bank (IADB), he was Chief of the Inclusive City Unit. In this role, he managed a $44.5 million project in the Latin America and Caribbean (LAC) region on developing 40 blended finance investments for small- and medium-sized enterprises in sustainable business models for urban service delivery. As the Lead Private Sector Specialist for PPPs, David led the IADB program to promote PPPs in Latin America and the Caribbean, providing technical assistance to governments to improve policy, project preparation, and implementation for sustainable infrastructure. These technical assistance projects are linked to $2 billion in investments in PPP projects. David developed a PPP Readiness Index entitled Infrascope, published by the Economist Intelligence Unit, which measures PPP institutional capacity in the LAC region, and was subsequently adapted by other multilateral development banks for Asia and Africa. Since leaving IADB, David has served as a PPP Consultant to the World Bank, Global Infrastructure Facility (GIF), and ADB, advising on infrastructure governance and preparing knowledge products on quality infrastructure investment, infrastructure governance, demonstrating results and value addition.

Abstract

This publication traces the evolution and experience of public–private partnerships (PPPs) in the urban water sector in Asia. It identifies levers and actions to enhance efficiency, value for money, and bridge service delivery gaps.

Using a broad PPP definition to cover contractual arrangements between public authorities and private operators, this report discusses an array of PPP structures, including concessions, leases, and management contracts. Also included is a discussion on design–build–operate (DBO) contracts and their variants, given their growing adoption in some regions of Asia.

The report builds on a review of macro-trends in water PPP transactions globally and a close examination of seven case studies from across the value chain, covering water distribution, water treatment, and wastewater treatment. It recognizes the challenging nature of implementing PPPs in the urban water sector, given that water's public good necessity and economic value characteristics remain weakly reconciled across much of developing Asia. It discusses the contextual diversity and various challenges faced by different regions in Asia in this regard.

Also underscored in this report is the importance of strengthening governance and public institutions by deepening sector reforms, building rigor in project preparation, ensuring competitive and transparent project bidding, and strengthening post-award project monitoring.

This publication highlights three central points, or pivots, on which governments need to act to scale urban water PPPs and their development impact: (i) water governance, (ii) enablers for PPPs, and (iii) critical elements of project-level transaction design.

Executive Summary

About This Publication

This publication builds on a review of macro-trends in water PPP transactions using data from ADB's *Asian Water Development Outlook* and the World Bank's Private Participation in Infrastructure (PPI) database. It analyzes seven select case studies in Asia from across the value chain, covering water distribution, water treatment, and wastewater treatment.

A broad PPP definition was adopted in this report to cover contractual arrangements between public authorities and private operators, embodying three principles: (i) the scope of definition covers both construction and *operations*, (ii) the contractual period lasts at least *5 years* post-construction, and (iii) *financing* is provided by either a private operator, a public authority, or both. Accordingly, this report covers an array of PPP structures, including *concessions, leases, and management contracts*. It also discusses *design–build–operate (DBO)* contracts and their variants through specific case studies, given their growing adoption in some regions of Asia.

The aim of this publication is to complement the body of research on urban water PPPs through a focus on governance and public sector-related factors, including *sector reform, project preparation, project bidding, and project monitoring*, which are necessary for effective PPP design and implementation. While the study does attempt to reflect the various underlying contextual differences and challenges faced by different regions of Asia when it comes to PPPs in the water sector, a detailed cross-country and regional comparison is beyond its scope.

Three Pivots to Scale Urban Water PPPs and Their Development Impact

Implementing PPPs in the urban water cycle (from water treatment and distribution to wastewater treatment and management) has been a challenge. With some exceptions, the characteristics of water as a public good necessity and the economic value of providing universal access remain weakly reconciled across much of developing Asia.

A syndrome of "low investment–poor service delivery–low-cost recovery" persists and undermines public water utilities institutionally and financially. Framed against this difficult backdrop, urban water PPPs are often set up to fail, and it is not surprising that they have not acquired a greater scale. Nevertheless, a review of PPP experience globally and in Asia confirms that when the preparation of PPPs identifies and allocates risks properly, accounts for life-cycle costs to optimize value for money in comparison with traditional public procurement, and remains fiscally sustainable, they can deliver positive outcomes, especially on operational efficiency dimensions. Many PPP cases from Asia that were reviewed in this study have reported positive operational access and financial performance outcomes. Building on these findings, this study identifies three complementary pivots along which governments need to act to scale urban water PPPs and their development impact.

The first is to put in place a holistic water governance framework that encompasses three actions:

(i)   an expressed public policy commitment to water security and inclusive access that is fiscally sustainable;
(ii)   empowered and capable public counterparty institutions mandated with service delivery, contract monitoring, and regulatory functions; and
(iii) a progressive revenue regime where revenues are linked to usage and service, and subsidies are transparent, targeted, and formalized in advance.

The second is to foster an enabling environment for PPPs comprising three aspects:

(i)   a sector-specific PPP strategy that identifies clear objectives and a pipeline of projects to aid programmatic implementation;
(ii)   greater rigor in project preparation and closure; and
(iii) transparent fiscal support rules and consistent frameworks to manage and monitor fiscal costs and contingent liabilities.

The third pivot is to design transactions in the project that incorporate features for bankability and competitive tension and focus on outcomes through the following:

(i)   supply-side impetus, competition efficiency, and transparency of procurement;
(ii)   a sharp focus on operations and maintenance (O&M), clear performance linkages, and post-award management;
(iii) contract sanctity and payment security; and
(iv)   contextual fit and appropriateness.

Key Messages

The unprecedented scale and pace of Asia's urbanization exacerbates its water deficits. The urban population of developing Asia grew fivefold from 375 million to 1.84 billion between 1970 and 2000, according to the *Asian Development Outlook 2019 Update: Fostering Growth and Inclusion in Asia's Cities.* This translates to an average annual growth of 3.4% and is much higher than the 2.6% growth in the rest of the developing world (Africa and Latin America) and the 1.0% in developed countries during this period. Developing Asia has also become more urbanized, with nearly half of its population living in cities in 2018, up from 20% in 1960.

This unprecedented urbanization sharply exposes Asia's water and sanitation access deficits. The *Asian Water Development Outlook 2020* of ADB observed that a staggering 2.1 billion people lack adequate access to water supply and sanitation. The International Institute for Applied Systems Analysis also notes in *Water Futures and Solutions: Asia 2050* that up to 1.4 billion people could be living in water-stressed regions of Asia and the Pacific by 2050. The 2017 ADB report *Meeting Asia's Infrastructure Needs* pegs Asia's investment requirements in the water sector at $800 billion between 2016 and 2030, or $53 billion annually, which is several times higher than the current investment.

Much of developing Asia has been slow to embrace PPPs in the water sector. According to the World Bank's Private Participation in Infrastructure (PPI) database, Asia implemented $12.6 billion worth of PPP project transactions between 2010 and 2019, with the People's Republic of China (PRC) accounting for close to 76% of this amount in value terms. The value of PPPs in the rest of Asia (about $3 billion) is relatively small in comparison with other developing regions. For instance, the Latin America and Caribbean region reported $17 billion during the same period. Much of Asia, therefore, seems to be lagging in the adoption of PPP in its water sector.

Treatment and upstream projects have dominated Asia's public–private partnership portfolio. A review of the sub-sectoral classification within the PPI database indicates that treatment plants accounted for nearly 83% of the value of PPPs in water in Asia between 2000 and 2019. This is in sharp contrast with trends from Latin America where PPPs in water utilities (or distribution) accounted for 78% of PPPs implemented by value. Therefore, even within the relatively small base of PPP projects, PPPs focused on distribution have been harder to implement in Asia. This is not surprising given the wider institutional and financing constraints in Asia's water sector.

Three constraints limit wider adoption of public–private partnerships in much of developing Asia. First, most Asian cities struggle to reconcile water's characteristics as a public good necessity and the economic cost of providing access. Water utilities across Asian cities rely excessively on fiscal transfers and have poor cost recovery (often less than their O&M cost obligations), low employee productivity, poor collection efficiency, and high nonrevenue water levels. Second, these policy and institutional weaknesses lead to a perpetual cycle of low investment–poor service delivery–low-cost recovery and weaken public water utilities institutionally and financially. Third, people resort to coping solutions such as bore wells and tanker supply, which leads to informal and unregulated privatization by neglect, and further weakens incentives for public service delivery. Framed against this difficult backdrop, even the few PPPs developed often become vulnerable to a "set up to fail" syndrome.

Experience globally and from select project cases in Asia suggests that water public–private partnerships can yield positive outcomes when effectively structured and implemented. This requires a proper governing framework for investment planning and fiscal decision-making and credible contract monitoring capacity. Nevertheless, the complexity of PPP preparation, structuring, and finance remains a challenge across much of the developing world. A 2009 World Bank report titled *Public–Private Partnerships for Urban Water Utilities: A Review of Experiences in Developing Countries* observed that by 2007, over 84% of 260 PPP contracts awarded since 1990 had remained active and served over 160 million urban residents in developing countries. This report also observed that nearly one-third of these projects, catering to over 50 million residents, were classified as *broadly successful*. Water PPP cases profiled for this study from various regions of Asia, including in Armenia (Central Asia), Malviya Nagar Delhi (South Asia), and Manila (Southeast Asia), also report the opportunities and challenges of attaining value for money, including in operational access and financial performance outcomes.

Public–private partnerships are beginning to make a mark in developing Asia; this is evident in diverse contractual arrangements and subsectors where PPPs have been implemented. A variety of models are emerging to maximize value-for-money benefits in various subsectors of the water value chain. Understandably, Asia's diversity is reflected in its multitude of different contractual arrangements with context-appropriate structuring.

Bulk water treatment and transmission projects: The PRC has seen a surge in water treatment projects. Indonesia is also implementing PPPs upstream, including the Umbulan Water Supply project, where construction is complete, and projects are under development in Sembarang, Jatilahur, and other locations. These projects are structured with strong government support mechanisms, including guarantee support from the Indonesia Infrastructure Guarantee Fund; viability gap financing and support for project preparation from the Ministry of Finance, Government of Indonesia; and offtake commitments from the city-level utilities to partly help offset and delink tariff risks.

Wastewater treatment: Similarly, PPPs are beginning to make their presence felt in the wastewater treatment subsector, which has long been within the realm of traditional public finance. India, for instance, has managed to implement several sewage treatment projects in cities along the river Ganges using hybrid annuity model concession contracts. Under these contracts, 40% of the project cost is paid upfront as compensation for construction, while the remaining 60% is paid over the life of the contract and linked to performance and service outcomes. While expeditious implementation of the network to feed these new plants and financial reforms at the level of utilities are critical, these projects have helped crowd-in private financing in an under-invested sector.

Water distribution: This subsector has seen a variety of contracts, including concessions (Nagpur, India; and Manila, Philippines), management or lease contracts (Yerevan, Armenia), and performance-based annuity contracts (Coimbatore, India). ADB has supported DBO contracts and performance-based construct and operate contracts in South Asia (Ilkal, India). The cases reviewed indicate a focus on bringing in private expertise for operational efficiency and service quality, with private sector financing of investment a secondary objective. While investment financed by the private sector is, indeed, prevalent, risks relating to water demand and tariffs have been largely off the table in water distribution PPPs in the last decade.

Mainstreaming PPPs: Scaling the development impact of PPPs requires concerted actions from governments concomitantly along three pivots. First, it is important to put in place a holistic water governance framework that encompasses the following three actions:

(i)   an expressed public policy commitment to water security and inclusive access;
(ii) empowered and capable public counterparty institutions mandated with delivery and regulation; and
(iii) a buoyant revenue regime and transparent targeted subsidies.

Second, governments need to foster an enabling environment for PPPs comprising the following three aspects:

(i)   a sector-specific PPP strategy that enunciates clear objectives and a pipeline of projects to aid programmatic implementation;
(ii) greater rigor in project preparation; and
(iii) fiscal support and frameworks to manage fiscal costs and contingent liabilities.

Third, transaction design at the project level should incorporate features for bankability, balanced risk allocation, and outcome focus through the following:

(i)   supply-side impetus and competition efficiency;
(ii) a sharp focus on operations and maintenance, clear performance linkages, and post-award management;
(iii) contract sanctity and payment security; and
(iv) contextual fit and appropriateness.

I. Asia's Urban Water and Sanitation Context

Rapid Urbanization Exacerbates Water and Sanitation Gaps

Asia has experienced a period of intensifying urbanization that could amplify urban water and sanitation deficits. Asia's urban population grew fivefold from 375 million to 1.84 billion between 1970 and 2017 for an annual average growth rate of 3.4%, driven by a period of strong economic growth.[1] The growth rate in Asia is much higher than the rates (2.6%) in the rest of the developing world (Africa and Latin America) and (1.0%) in developed countries during this period. Developing Asia has also become more urbanized; nearly half of its population lived in cities in 2018, up from 20% in 1960. According to the United Nations Human Settlements Programme (UN-Habitat), Asian cities are also among the most vulnerable in the world to climate hazards, especially for informal settlements located in environmentally fragile areas.[2]

Access Deficits and Investment Needs are Sizable

A rapidly urbanizing Asia faces severe deficits in water and sanitation, with transition to universal access remaining a challenge. In 2017, an estimated 300 million people lacked access to safe drinking water and 1.5 billion lacked basic sanitation.[3] The challenges in sanitation are particularly acute. An Asian Development Bank Institute (ADBI) paper notes that 780 million people practice open defecation, and 80% of wastewater lacks appropriate treatment.[4]

Over 57% of the urban population do not have access to toilets, including containment, treatment, and end use treatment and disposal (see Table 1). Even where basic infrastructure for water and sanitation services exists, investments are needed to improve services, such as continuous piped water supply and hygienic conditions (including hand hygiene), which have gained more attention from policy makers since the outbreak of the COVID-19 pandemic.

The *Asian Water Development Outlook (AWDO) 2020* of ADB assessed urban water security as Key Dimension 3 (KD3) covering water supply, sanitation, affordability, drainage, and environmental water security.[5] The AWDO analyzes the status of water security in Asia and the Pacific. The publication defines water security as access to a safe and affordable water supply, sanitation, livelihoods, and healthy ecosystems that lower water-related risks. It presents a framework for water security with five key dimensions (KDs)—(i) rural household water, (ii) economic

1   ADB. 2019. *ADO 2019 Update: Fostering Growth and Inclusion in Asia's Cities.* September. Manila.
2   United Nations Human Settlements Programme (UN-Habitat). 2010. *The State of Asian Cities 2010/11.* Japan. p. 21.
3   ADB, Manila. 2017. *Meeting Asia's Investment Needs.*
4   ADB Institute and Institute of Water Policy, LKYPP Singapore. 2019. *Water Insecurity and Sanitation in Asia.*
5   ADB. 2020. Asian Water Development Outlook 2020. Manila

water security, (iii) urban water security, (iv) environmental water security, and (v) water-related disaster security. The 2020 report gave a composite score of 13.3 out of 20 for developing Asia and the Pacific. All subregions, except for East Asia (which stands out due to the improved sanitation coverage of the PRC), exhibit severe gaps (see Table 2).

The investments needed to bridge these massive legacy gaps and provide for the expected growth in the urban population are, therefore, sizable. An ADB study estimated Asia's water sector investment needs at $787 billion between 2016 and 2030.[6] Yet, in most of developing Asia, actual investments in water and sanitation trail significantly and require a sharp scale-up from current levels (Table 3).

Table 1: Water, Sanitation, and Hygiene Access in Asia
(estimated coverage, %)

Type		Region	West Asia	Central Asia	South Asia	Southeast Asia	East Asia
Water	**Basic**	Urban	96	93	85	70	98
		Rural	79	81	83	81	87
	Safely managed	Urban	84	75	48	46	90
		Rural	54	20	9	12	45
Sanitation	**ODF**	Rural	100	100	51	84	98
	Basic	Urban	94	95	65	81	86
		Rural	76	84	33	65	64
	Safely managed	Urban	41	35	6	37	24
		Rural	34	34	34	34	34
Hygiene	**Hand washing**	Urban	97	92	85	93	83
		Rural	92	77	49	79	44

ODF = open defecation free.

Source: World Bank. 2016. *The Costs of Meeting the 2030 SDG targets in Water, Sanitation, and Hygiene.* Water and Sanitation Program.

Table 2: Population-Weighted Average Key Dimension 3 Results, 2020

Region as per ADB Classification	Water Supply	Sanitation	Affordability	Drainage	Environment	KD3 Score
Central and West Asia	3.6	1.9	4.8	0.8	0.4	**11.5**
East Asia	5.9	5.9	4.7	0.9	0.6	**17.9**
Pacific	1.4	1.3	1.6	1.0	0.9	**6.3**
South Asia	2.7	1.2	4.7	0.9	0.6	**10.1**
Southeast Asia	3.0	2.0	5.3	0.8	0.9	**11.9**
Advanced Economies	5.9	5.9	4.8	0.9	1.0	**18.4**
Asia and the Pacific*	3.9	3.1	4.8	0.9	0.6	**13.3**

* Without advanced economies.

Note: Maximum score for KD3 is 20; numbers may not sum precisely because of rounding.

Source: ADB. 2020. *Asian Water Development Outlook 2020.* Manila.

---

[6]   ADB. 2017. *Meeting Asia's Infrastructure Needs.*

Table 3: Infrastructure Investment in Asia and the Pacific by Sector, 2016–2030
($ million)

	Investment Needs	Annual Average	Share of Total (%)
Power	11,689	779	51.8
Transport	7,796	520	34.6
Telecommunications	2,279	152	10.1
Water and sanitation	**787**	**52**	**3.5**
Total	**22,551**	**1,503**	**100**

Source: ADB. 2017. *Meeting Asia's Infrastructure Needs*. Manila.

SDG 6 aims to ensure clean water and sanitation for all by 2030.[7] This includes (i) universal and equitable access to safe and affordable drinking water and (ii) adequate and equitable sanitation and hygiene for all, ending open defecation, especially for women and girls. A technical paper by the World Bank's Water and Sanitation program pegs at $55.5 billion per year the investment needed to achieve SDG 6 and improve water, sanitation, and hygiene conditions for Asia's developing economies between 2016 and 2030. This includes urban investment needs of $35 billion annually (Table 4).

Table 4: Investments Needed in Asia and the Pacific for Sustainable
Development Goals 6.1 and 6.2 on Water, Sanitation, and Hygiene
($ million)

Region	West Asia	South Asia	Southeast Asia	East Asia	Asia Total
Urban	3,414	11,105	7,181	13,392	35,092
Rural	1,293	13,417	3,237	2,498	20,445
Total	4,707	24,522	10,418	15,890	55,537

Source: World Bank. 2016. *The Costs of Meeting the 2030 Sustainable Development Goal Targets on Drinking Water, Sanitation, and Hygiene.* Technical Paper, Water and Sanitation Program. Washington, DC.

This sizable investment challenge is compounded by the impact of intensifying climate change and the need for infrastructure resilience. It is also amplified by weak governance structures and institutional capacity limitations in several developing economies in Asia. Public institutional capacities to plan, develop, finance, and manage infrastructure investments for delivery of urban water and sanitation service remain weak. In this context, a judicious, planned, and well-regulated approach to secure private participation is seen as potentially complementary to efforts to drive accelerated improvements to service delivery in water and sanitation. Nevertheless, the role of the private sector in Asia's water services has been rather limited.

The following sections of the publication trace the evolution of PPPs in the urban water sector in Asia, distil lessons and insights from case studies of select PPP projects in Asia, and identify levers and actions for a scaled-up adoption of PPPs to bridge service delivery gaps in Asia's urban water and sanitation sector.

---

[7]   See more information on SDG 6 and related targets at https://data.unicef.org/sdgs/goal-6-clean-water-sanitation/.

II. Emerging Trends from Asia's Water Public–Private Partnerships

This section traces the evolution of Asia's urban water PPPs from a qualitative perspective. It analyzes the trends in water PPP transactions identified in the World Bank's PPI database. Most of the PPP transactions in Asia have been recorded since the year 2000; therefore, this analysis of transaction history focuses on 2000-2019.

Evolution of Public–Private Partnerships in Asia: A Qualitative Assessment

There has been rise and fall in the scale of PPP transactions alongside changes in the overall policy context of various countries. Over the last two decades, the PRC has seen a meteoric rise in the volume and value of PPP transactions, becoming the largest water PPP market globally. Countries in Southeast Asia, like Malaysia and Thailand, have adopted sector-level actions to spur PPPs. Large concessions have been awarded in Manila, Philippines (1997), and Jakarta, Indonesia (1998). In Central Asia, Armenia initiated a range of policy reforms and awarded PPPs through lease and management contracts for Yerevan and adjoining regions from 2000 to 2005.[8] In South Asia, India awarded several water sector PPPs in the second half of the 2000s, which coincided with the beginning of the Jawaharlal Nehru National Urban Renewal mission (JNNURM), a scaled-up federal grant program. By 2015, most Asian countries had made efforts to create enabling environments for PPPs in the water sector.

Notwithstanding these developments, water PPPs in Asia have been vulnerable to tepid political commitment, project failures, and slow progress in sector reforms. Malaysia and Thailand have seen policy reversals in the last decade. The momentum of PPP in India sharply fell in the early half of the 2010s, although a revival is underway. Apart from Manila's city-wide concessions, PPPs in the Philippines have been few and far between. Indonesia's water PPPs have focused more on upstream treatment and transmission segments.

In Central Asia, Georgia enacted its PPP law in 2016 and followed it up with secondary legislation and the establishment of a PPP unit in 2018.[9] In Uzbekistan, a 7-year water supply PPP contract was awarded to Suez in July 2020. The PRC has seen the sharpest growth in adoption of PPPs in the last two decades. Other than the PRC and, to some extent, the Republic of Korea and Singapore, water PPPs in the rest of Asia have tended to follow a rather nonlinear trend in the last two decades.

---

[8]   Refer to case study on Yerevan water supply in Section 3 for additional details.
[9]   ADB Institute. 2020. *Public Private Partnerships in Georgia and Impact Assessment on Infrastructure.*

Cost recovery levels in Asia's public water utilities are low and often an inhibitor for PPPs. Among Asia's subregions, Southeast Asia tends to perform more strongly. An ADB study in 2014 found the average operating ratio (expenditure to income ratio) in South Asian cities to be a healthy 0.67 but observed collection efficiencies to be very low.[10] Cities in Central Asia often report cost recovery levels below O&M costs; the annual *Global Water Intelligence Survey* puts Georgia, Uzbekistan, and the Kyrgyz Republic in the bottom 15 of 186 countries on water tariffs.[11]

The adoption of PPPs in Asia is also constrained by a lack of institutional maturity and evolution of regulatory structures when compared with Latin America, where PPPs have been underpinned by well-developed regulatory structures (e.g., Chile). In Asia, contract regulation and tariff setting are often the domain of local governments, which may not have either the functional capacity or financial powers to manage large PPP projects. Outside Manila, where the Manila Water Supply and Sewerage Regulatory Office regulates PPP contracts, regulatory frameworks tend to be weak. Furthermore, distrust and skepticism of PPPs in water tend to be higher among citizen groups due to lack of effective institutional capabilities in many public water utilities and local governments. These capacity limitations and weak policy frameworks adversely affect the risk profiles of PPP projects in Asia.

Trends from Public–Private Partnership Transaction History of the Last Two Decades

The PRC and the Latin America and Caribbean region (LAC), taken together, had an 82% share by volume of PPP projects in the water sector and a 67% share by value in the last two decades. The rest of Asia accounted for a miniscule 10% share by volume and a 19% share by value. In almost all regions, the number of new project awards fell sharply between 2010 and 2019 in relation to the previous decade (2000–2009) (Table 5). While PRC and LAC showed growth across these periods in value terms, the rest of Asia witnessed a decline in both volume and value.

Within Asia, the PRC accounted for the lion's share of PPP project transactions in both value and volume. From 2000 to 2019, the PRC accounted for 85% of the volume and 60% of the value of water PPPs. In terms of value, the PRC was trailed by Malaysia (22%), India (4%), Philippines (4%), and Indonesia (3%). The PRC also accounts for the highest value of projects globally in this period. The PRC showed a steadier and consistent trend in project starts during the period compared with the rest of Asia, where the project starts showed variable trends (see Table 6 and Figure 1).

An overwhelming majority of PPP contracts in Asia between 2000 and 2019 were in treatment plants (524 of 638 projects awarded). Water distribution accounted for 108 projects, while six projects had water distribution and treatment components. Only about 10% of projects in the PRC were water distribution projects. Although the rest of Asia had 63% of its projects in water treatment, this is still lower than that in LAC, Europe, and sub-Saharan Africa (see Figure 2).

---

[10]   A.C. McIntosh. 2016. *Urban Water Supply and Sanitation in Southeast Asia: A Guide to Good Practice.* Manila.
[11]   GWI Magazine. 2020. *Global Water Intelligence.* September. United Kingdom.

Table 5: Water Public–Private Partnerships Awarded Globally by Region, Volume, and Value

	2000–2009				2010–2019				2000–2019			
	No. of Projects	%	$ million	%	No. of Projects	%	$ million	%	No. of Projects	%	$ million	%
Asia	378	64	16,860	55	260	71	12,639	40	638	68	29,500	47
PRC	312	53	8,013	26	234	64	9,615	30	546	58	17,629	28
Rest of Asia	66	11	8,847	29	26	7	3,024	10	92	10	11,871	19
LAC	134	23	7,870	26	90	25	16,334	52	224	24	24,203	39
SSA	15	3	164	1	4	1	219	1	19	2	383	1
MENA	17	3	3,202	11	8	2	1,147	4	25	3	4,349	7
Europe	40	7	2,296	8	3	1	1,333	4	43	5	3,630	6
Total	**584**	**100**	**30,392**	**100**	**365**	**100**	**31,672**	**100**	**949**	**100**	**62,064**	**100%**

LAC = Latin America and Caribbean, MENA = Middle East and North Africa, PRC = People's Republic of China, SSA = Sub Saharan Africa.

Source: World Bank PPI database.

Table 6: Public–Private Partnerships in Asia by Volume and Value
($ million)

	Volume		Value	
	Nos.	%	$ Million	%
PRC	546	86	17,629	60
Malaysia	11	2	6,502	22
India	20	3	1,258	4
Philippines	8	1	1,249	4
Indonesia	11	2	759	3
Thailand	16	3	544	2
Others	26	4	1,560	5

PRC = People's Republic of China.

Source: World Bank PPI database.

Of the top five projects between 2010 and 2019, three were from the PRC, one from India, and one from the Philippines:

1. Changshu Sino French Water, PRC $517 million
2. Coimbatore Distribution Contract, India $497 million
3. Bulacan Bulk Water Supply Project, Philippines $489 million
4. Chengdong Wastewater Treatment Plant, PRC $485 million
5. Dali Eryuan County Urban and Rural Wastewater Utilities Project, PRC $348 million

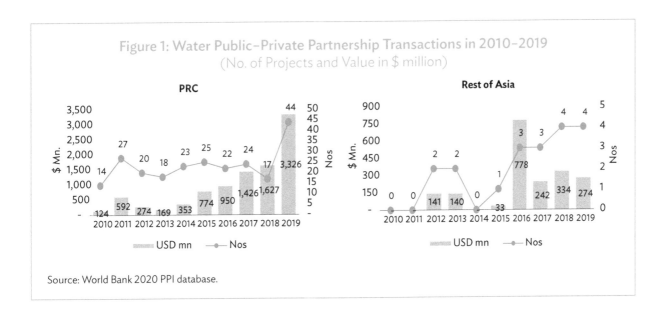

Figure 1: Water Public–Private Partnership Transactions in 2010–2019
(No. of Projects and Value in $ million)

Source: World Bank 2020 PPI database.

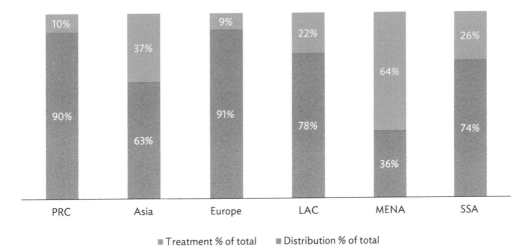

Figure 2: Share of Water Public–Private Partnerships by Type and Percentage of Total

MENA= Middle East and North Africa, PRC = People's Republic of China, SSA= Sub-Saharan Africa.

Note: In this figure, the PRC is shown separately given its dominance in water public–private partnerships (PPPs), and it is also included in the data for Asia.

Source: World Bank PPI database.

III. Insights from Select Public–Private Partnership Cases

This section examines select water PPP cases from Asia to obtain insights at the project level about what has worked and what has not. The projects that have been selected highlight the diverse approaches that different cities have taken to water governance, project structuring, and financial viability.

The diversity of starting points shows the complexity and challenges of structuring water sector PPPs. A "one size fits all" approach rarely works; tailoring PPPs to the specific context is critical. The seven cases analyzed in this section cover three categories: water distribution (five projects), water treatment and transmission (one project), and sewage treatment (one project). The number of water distribution projects reviewed is higher due to the heterogeneity of the models (see Table 7 for a summary of the project cases).

Water Distribution

Distribution has been an important segment for water PPPs given the imperatives to improve access and operating efficiency, thereby reducing water losses. At the same time, water distribution PPPs are difficult to structure due to resistance to user charge increases, financial sustainability pressures, and political reluctance. Furthermore, the projects are complex in view of their brownfield nature, underground assets, unavailable databases, and often unclear baseline performance. A World Bank report on PPPs in water distribution observed that by 2007, over 84% of 260 PPP contracts awarded since 1990 remained active and served over 160 million urban residents in developing countries. Projects catering to over 50 million residents (nearly a third of these projects) were classified as broadly successful.[12]

The distribution subsegment reflects highly diverse PPP contracting structures and models. These include city-level concessions in Manila and Jakarta in the mid-1990s; management or lease contracts in Armenia and Malviya Nagar, Delhi, India; staggered annuity contracts such as in Coimbatore, India; and investment-light DBO and performance-based construct and operate contracts (PBCOCs) in Mangalore, India. Case studies reflecting this heterogeneity are discussed below. Water distribution contracts are largely aimed at realizing operating efficiency gains. ADB has been supporting DBO contracts and PBCOCs in South Asia.

Realizing operational efficiency and expanding access seem to be more critical objectives in this segment than the level of private sector investment, which is increasingly becoming a secondary objective. The case studies seem to substantiate that even though private sector financing is prevalent, the risks relating to water demand and tariffs are not addressed in many of the newer PPP contracts in recent years; management contracts and leases have become the preferred mode of PPP contracting over investment concessions.

---

[12] World Bank Group. 2018. *Public–Private Partnerships for Urban Water Utilities.*

Table 7: Summary Profiles of Public–Private Partnership Projects Reviewed

	Yerevan, Armenia	Coimbatore, India	Jakarta, Indonesia	Malviya Nagar, Delhi, India	Ilkal, Karnataka, India	Umbulan, Indonesia	Cities along Ganges, India
Subsector	Water Distribution	Water Distribution	Water Distribution	Water Distribution	Water Distribution	Transmission	Sewage Treatment
PPP model	Management contract/ Lease contracts	Performance-based annuity BOT	Concession agreement for construction, O&M	Management Contract	Performance-based construct and operate contract (PBCOC)	BOT Concession	Hybrid Annuity Concession
Government Authority	Public Water Utility Armenia Water and Sewerage Company (AWSC)	Urban Local Body Coimbatore Municipal Corporation	Local Authority PAM Jaya	Public Water Utility Delhi Jal Board	Karnataka Urban Infrastructure Dev Corporation	Public Water Utility PDAM	State Water Utilities Jal Nigam
Year signed/ duration/ status	2004 / 13 years Phase I completed	2017 / 26 years Construction phase	1998 / 20 years Terminated	2013 / 12 years Operations	Construction (2 years) Operation (4 years)	2017 / 25 years Operations	2017 / 25 years Construction
Population served	0.7 million	1.6 million	10 million	0.382 million	0.051 million	1.3 million	4.35 million
Bidding / bid variable	Competitive Lowest annual fee	Competitive Net present value of cost quoted	Unsolicited	Gross water rate per m^3	Lowest bid price	Lowest fiscal incentive	Lowest bid value
Operator	Saur Group France	Suez Environment India	PALYJA, Suez	Suez Environment India	Veolia India	Consortium of PT Medco Energi International Tbk. and PT Bangun Cipta Kontraktor	HNB engineers (Haridwar) Essel Infra Projects Ltd (Varanasi) Triveni Engineering & Industries Ltd (Mathura)
Private investment	NIL	30%	Not available	30%	No investment by operator	40%	Nil
Government grant	100%	70%	Not available	70%	100%	60%	40% (as EPC payment)
Revenue model	Fixed fees + performance	Tariff + annuity payments	Tariff + annuity payments from government	Gross rate * volume billed	Fixed (60%) and performance (40%)	Annuity payments from	Initial project cost + annuity payments

Source: Secondary research by the authors.

Yerevan Water Supply, Armenia: Management or Lease Contracts

Yerevan's water supply has experienced three generations of public private partnership (PPP) contracts in the last two decades.[13] The first was a PPP management contract in 2000, which was followed by second- and third-generation contracts in 2006 and 2017. The outcomes reinforce the positive impacts of staying the course with PPPs, especially when backed with commitment to carry out enabling policy and sector-level reforms. A phased shift to deeper private engagement, building on lessons from earlier phases, along with enabling policy actions to strengthen institutions and reform tariffs, and complementary multilateral financing to meet investment needs have delivered tangible development impacts and improvements in service delivery.

Drivers underlying the public–private partnership contract

Armenia had been experiencing a deterioration of its water networks since its independence in 1991. The water supply coverage of Armenia is relatively good; however, less than 15% of water connections received a continuous supply. Metering was not very prevalent and nonrevenue water levels were high. With user charges limited due to flat tariffs, O&M cost recovery was less than 30%. To address these challenges, the government started its PPP program in 2000 to contract private operators in the management of Armenian water operations.

Armenia had different first-generation water supply PPPs between 2000 and 2016. The first PPP contract was a management contract (2000–2005) awarded for Yerevan (with a population of 1.2 million) by Yerevan Water Services Company (YWSC), the city's public water utility. This was followed by a second PPP contract by Armenia Water Services Company (AWSC) for 29 towns and 160 villages outside Yerevan (with a combined population of 620,000). The third PPP contract was awarded for three smaller water service companies in Lori (LWSC), Shirak (SWSC), and Nor Akunq (NAWSC); the three had a combined population of 320,000.

In 2017, a combined national-level second-generation PPP lease contract was awarded to cover water and sewerage services in all regions (see Table 8). This case study focuses on the PPP contracts awarded under the YWSC.

Even in the absence of national PPP policies or legislation, the PPP contracts were accompanied by a range of sector reforms. In 2002, the law on local self-government delegated the responsibility for providing sewerage and water supply services to local authorities. The State Committee on Water Systems (SCWS) was established the same year, and the Water Code was formulated to provide an overarching legal framework for managing water resources in Armenia. The National Water Policy of 2005 and the National Water Program of 2006 followed.

The Public Services Regulatory Commission (PSRC) was established to regulate public utilities, and the Yerevan and Armenia water utilities, YWSC and AWSC, were formed. The Government of Armenia concurrently mobilized funding from multilateral agencies, including the World Bank and ADB. Immediately upon becoming a member in 2006, the Government of Armenia enlisted ADB support in water sector reforms. ADB extended its support in 2007 to expand, upgrade, and improve infrastructure and service delivery. These efforts to mobilize funds, develop enabling policies, and reform institutions were critical to engaging private sector participation and improving service delivery, even though Armenia did not yet have a PPP policy or related legislation.

---

[13]   This case has been prepared with inputs from the following source documents: (1) ADB. 2011. *Armenia Water Supply and Sanitation. Challenges, Achievements, and Future Directions.* Manila, (2) ADB. 2019b. *Project Performance Evaluation Report for Loans 2363 and 2860-ARM: Water Supply and Sanitation Sector Project.* Manila, (3) World Bank. 2018a. *Armenia Municipal Water Project—Project Performance Assessment Report.* Independent Evaluation Group, and (4) ADB. 2018. *Project Completion Report Armenia: Water Supply and Sanitation Sector Project.* Manila. Inputs from interviews with ADB's Central and West Asia team have also been incorporated.

Table 8: Water Distribution in Armenia: Summary of Contracts Awarded

Water Board	Contract Type	Contract Period	Operator	Population Served (Million)	Bidding Criteria	Bids	Investment ($ Million)
YWSC	1. Management	2000–2005	Acea	1.03	Lowest fees	3	28
	2. Lease	2006–2016	Veolia		Lowest tariff	--	68
AWSC	3. Management	2004–2016	Saur Worldwide	0.62	Lowest annual fees	4	180
SWSC LWSC NAWSC	4. Management	2008–2016	MVV Decon, MVV Energy, AEG Services	0.32	Lowest annual fees	3	74
All	5. Lease	2017–2032	Veolia	2.2	Lowest tariff	4	628

Acea = Azienda Comunale Elettricità e Acque.

Note: SWSC, LWSC, and NAWSC are water and sewerage companies of Shirak, Lori, and Nor, respectively.

Source: Secondary research.

Salient features of the public–private partnership arrangements

Management contract (2000–2005): Following a 2-year project preparation period, funding mobilization, and competitive bidding, the Italian utility Acea was selected from among the three bids received. Acea took over O&M of Yerevan's network in June 2000. The contract included 93 key performance indicators (KPIs). Four of them were linked to bonuses: (i) meter installation, (ii) survey of leak detection, (iii) use of electricity, and (iv) water supply continuity.

Lease contract (2006–2016): The lease contract (2006–2016) was revised to facilitate increased risk transfer to the private sector, including billing and collection risks. The contract gave the responsibility for investment implementation to the private operator to steer the network rollout for greater efficiency. The operator was required to finance a part of the capital expenditure (capex) but was given flexibility for implementation. The operator was also required to provide a leasing fee to the State Committee for Water Economy (SCWE). This was used to service debt repayments.

The service area was expanded to cover 30 villages around Yerevan. For better focus, the number of KPIs was reduced from 93 to 25. Four KPIs were subject to penalties: continuity, quality, speed of responsiveness to complaints by customers, and compliance with time limits for implementation of capex. The tariff increased significantly after the start of the contract to AMD 173/m^3 ($0.39/m^3), up from a prelease tariff of AMD 125/m^3 ($0.28/m^3). A 10-year program for tariff levels, with rules for tariff adjustment procedures, was established. Tariff regulation was performed by the PSRC based on submissions by the operator and the approval of SCWE. In 2009, the private operator and SCWE renegotiated the KPIs based on experience from the early years.

The negotiations lowered target hours of supply from the third to seventh year of the contract, increased target hours for eight to ninth year, and left them the same for the final year.[14] The lease contract improved both operational and financial results in the performance of YWSC. Water services improved sharply; however, NRW levels remained high.

---

[14]   P. Marin, D. Muzenda, A. Andreasyan. 2017. *Review of Armenia's Experience with Water Public–Private Partnerships.* Washington, DC.

Outcomes and the way ahead

The lease contract led to positive operational and financial improvements. Increased tariffs and collection rates led to a sharp increase in revenue. Water services became fully self-financing by 2011, although nonrevenue water (NRW) levels remained high. From 2006 to 2015, the operator recorded approximately $4.1 million in cumulative operational profit before taxes. Table 9 shows the project outcomes on key performance metrics.

Table 9: Outcomes under Two Phases of the Yerevan Water Distribution
Public–Private Partnership

Parameter	Phase I Yerevan Management Contract (2000–2005)	Phase II Yerevan Lease Contract (2006–2016)
Water supply continuity	From 4 to 18 h per day	From 18 to 23 h per day
Electricity consumption	Decreased by 48%	Decreased by 82%
Water losses	–	83%–75%
Bill collection rate	20%–80%	80%–97%
Metering % of connections	7–63	87–98
Tariff change	1999: AMD 56/m³ ($0.10) 2005: AMD 125/m³ ($0.27) 170% increase	2006: AMD 125/m³ ($0.28) 2016: AMD 170/m³ ($0.35) 36% increase
Operating profit / (loss)	Loss ($19.5 million)	Profit $9.6 million (after year 5)
Capex	Capex: $28 million ($4.7 per capita/year)	Capex: $68 million ($5.7 per capita/year)

Source: Secondary research by authors.

Multilateral support in the form of both funding and technical assistance has been a key enabler for the program. The World Bank and ADB supported the investment program and assisted the government in implementing several reform measures related to preparation of tender documents, tariff reforms, achievement of operational efficiency, and improvements in financial sustainability. The European Bank for Reconstruction and Development and the EU also played critical roles. Kreditanstalt für Wiederaufbau (KfW) was also involved in regional utilities from 2009 to 2016, and recently supported water supply facilities in 30 villages. The project provides some interesting lessons and replicable practices for structuring PPPs in water distribution.

a. **Political stewardship, leadership stability, and commitment to reform:** The commitment of the Government of Armenia to reforms and institutional stability was a key factor. SCWE was a strong and supportive public counterpart and provided stable leadership, with the same chairperson leading SCWE through most of the first generation of PPPs. This was vital in creating an atmosphere of trust with the private operators. Even as it embarked on the PPP program, the government implemented a series of supporting reforms through the various phases. For instance, enabling legislation on metering and revenue collection and the creation of empowered institutions in the form of SCWS and PSRC were critical to strengthening public sector capacity and accountability.

b. **Sustained commitment to public–private partnerships while incorporating lessons from experience:** PPPs were introduced on a city-wide basis and expanded to other regions in a phased manner. Every time a new contract was signed, there were improvements over the earlier contracts in the structure, KPIs, and focus areas for service delivery. For instance, the second contract in Yerevan moved to a lease contract structure with higher risk transfer and better focus on cost recovery, while the latest contract has introduced KPI and performance obligations with respect to NRW reduction.

c. **PPP focus around service delivery rather than investment financing:** Although later contracts envisaged some level of private financing, Armenia seems to have clearly settled toward engaging the private sector in service delivery improvement, complementing private experience with concessional multilateral financing. Technical assistance, including from ADB and the World Bank for policy and tariff reforms, project preparation, and driving midcourse corrections through periodic evaluations have also contributed immensely.

d. **Taking a country-wide approach for efficient water supply building on lessons from earlier contracts:** In 2016, Armenia opted for a country-wide approach, including moving from multiple PPP contracts to a composite contract with Veolia for 2017-2032. This is relatively rare and ought to be a subject of further study and research. As the focus of this report is to take a view of the overall chronology of the reform, a detailed evaluation of this new country-wide approach is beyond its scope. Nevertheless, there is value for policy makers in evaluating the recent developments and impact of the country-wide approach that Armenia seems to have adopted.

Malviya Nagar, Delhi, India: Performance Management Contract

Malviya Nagar is a 14 km^2 command area within the national capital region of Delhi. The Malviya Nagar performance management contract has shown strong improvements in operational and financial performance within its first 7 years. These encouraging results deserve attention for possible replication at scale within Delhi, in other large metropolitan cities in India, and in other developing economies.

Drivers underlying the PPP contract

The Delhi Jal Board (DJB) is the public water utility that manages water supply in the National Capital Territory of Delhi. It manages an 11,000 km water supply network and services 1.7 million connections for a population of 17.5 million over an area of 1,500 sq. km. In 2011, an empowered committee, set up by Delhi's Chief Minister to identify areas for management reforms and water distribution, endorsed DJB's proposals to award PPP contracts in two localities in South Delhi: Malviya Nagar and Vasant Vihar. Malviya Nagar is a densely populated locality in South Delhi spread over 0.9% of Delhi's area with a population of around 400,000 (2.3% of Delhi's population). After competitive bidding, the PPP contract for Malviya Nagar was awarded to Suez India in October 2012 and commenced on 1 January 2013.

Salient features of the public–private partnership arrangement

The project was to be executed through a 12-year performance management contract. It envisaged a 2-year period for study and construction, followed by a 10-year O&M period.

a. The DJB was to invest 70% of the initial network capex of ~Rs236 crore ($36 million), and the operator was to invest the rest (30%). The operator was also responsible for O&M of the network and the full customer cycle (meter reading, billing, collection, and complaint handling). Bulk water was made available from the main reservoir of DJB servicing Malviya Nagar. The objectives of the contract were the following:

   (i)   move to continuous supply in 3 years while increasing the daily duration of supply from 3 h to 8 h,

   (ii)   decrease NRW from 67% to 15% in 12 years, and

   (iii) provide full coverage and improved customer management.

The project involved construction and rehabilitation of 100 km of the existing 200 km network and replacement of all water meters. The operator was to be compensated based on the bid value of gross rate per kiloliter (kl) for the volume of water supplied. User charges collected by the operator at the prevailing rates were to be deposited into an escrow account. The contract provided for readjustment of the gross rate in case of changes in business plan assumptions.

b. Eight KPIs were monitored under the contract, and the operator's compensation was linked to these through penalties and revenue impact. The KPIs included: (i) increasing coverage of water supply from 84% to 100%, (ii) reducing per capita supply of water from 286 liters per capita per day (LPCD) to 190 LPCD, (iii) ensuring continuity of supply from 3–8 h to 24 h, (iv) increasing metering from 41% to 100%, (v) bringing the quality of water to 100%, (vi) redressing 80% of complaints, (vii) bringing user charges collection efficiency to 95%, and (viii) keeping NRW below 15%.

c. The operator and DJB had to work together to resolve several technical, operational, commercial, and communication challenges early in the contract. Network maps and asset information provided to the operator were insufficient and, in some cases, inaccurate. In several cases, the actual quantities in construction varied from the bill of quantities (BOQs) provided. Network surveys and construction were hampered by the presence of multiple underground utilities, often in narrow congested streets and lanes, and a lack of knowledge of the network specifics among DJB staff. Residents initially viewed the project with suspicion, distrust, and concern over tariff increases. Resistance also came from sizable unauthorized settlements in the service area that were not captured under the connections and tariff net. The operator faced challenges of delayed payment or nonpayment early on. O&M payments were also not made as required by the contractual obligations.

Outcomes and the way ahead

Notwithstanding these challenges, the operator, with support from DJB, managed to initiate and complete critical interventions early in the contract period. The operator undertook a geographic information system-enabled network mapping exercise and clean-up of the customer database and deployed a real-time operational performance system. Customer complaints were used to troubleshoot technical and network problems. Helium-based leak detection was deployed widely to improve network performance. The operator also put in place a communication outreach program for wider stakeholder acceptance and contact centers for customer management.

Early performance improvements in the first couple of years helped build wider stakeholder confidence. NRW was reduced from 68% to 46% within 18 months, while billed volume increased by 69% within the first 2 years. Although continuous 24/7 supply was demonstrated in some areas, it was rolled back to 16 h to reduce water wastage. Over 3,000 leaks on the network were successfully detected and repaired, and over 8,400 meters were replaced. By early 2020, over 100 km of the old network and 95% of the faulty house service connections had been replaced. Nearly 100 tube wells were decommissioned and over 17,000 billing disputes had been resolved. Mapping of 100% of customers and the network has been completed (see Table 10).

The positive performance outcomes, as reported above, should encourage DJB to replicate such arrangements on a scaled-up city-wide basis. In September 2020, the Government of Delhi announced a "One Zone One Operator" policy under which the operation and maintenance of water supply and sewerage lines in seven to eight zones will be handed over to private operators for 10 years or more, even as the government retains the overall responsibility for water procurement and supply.

Table 10: Reported Performance Improvements under Malviya Nagar
Water Supply Public–Private Partnership

Parameter	2012	2019
Population served	380,000	441,000
Connections (metered connections % of total)	32,148 (41)	48,067 (98)
Volume supplied MLD (Per capita supply LPCD)	75.79	66.12
NRW % (NRW in areas with 24/7 water supply %)	68% (45)	35% (11)
Billing Rs. Crore (Realization Rs./Kl)	7.6 (23.9)	61.6 (43.0)
Collection efficiency %	81	95
Coverage of supply %	84	100
Time for new connection (days)	62	14
Timely redressal % of complaints	–	86
Hours of supply (h)	3	3–24

Source: Presentations shared by Suez India (project operator).

Enablers and lessons learned

Even though Delhi has been attempting to initiate private participation in water supply since the early 2000s, its attempts had been stymied for various reasons, including large-scale protests in 2006. In the face of such setbacks, the approach to piloting performance-based contracts in smaller areas (but of scale) to drive technical, operational, and financial improvements and build confidence for wider replication appears to be a logical starting point.

The project reinforces that rapid efficiency gains and performance improvements in water distribution are, indeed, possible even in difficult brownfield terrain, when accompanied with policy commitment, political stewardship, and public sector capacity for monitoring. Clear performance-based contracting, with monitorable KPIs linked to compensation through incentives and penalties, also helps in driving outcome orientation. Given the teething challenges and legacy constraints imposed by brownfield projects, strong leadership and a longer stabilization and construction timeframe can also help set up such projects for success.

Coimbatore, India: Staggered Annuity Performance Management Contract

The Coimbatore Water Supply project is a brownfield project implemented through a performance-based management contract. The operator is expected to optimize, rehabilitate, and operate the water distribution system in an area of over 100 sq. km with a distribution network of 1,200 km and 150,000 connections. The operator has committed to improving customer services by implementing a state-of-the-art customer call center and customer agencies for resolving complaints and providing personalized services. In the early construction stages, some of the contract features, including its staggered annuity compensation scheme and KPI design, merit attention.

Drivers underlying the public–private partnership contract

Coimbatore is a large industrial city located in the south Indian state of Tamil Nadu, about 350 km from Bengaluru and 500 km from Chennai. The project was originally approved for funding under the Government of India's JNNURM program, wherein 50% of the funding for capex was to come from GoI, 20% from the State Government of Tamil Nadu (GoTN), and the rest from the private operator.

Although the project was approved by the Central Sanctioning and Monitoring committee under the JNNURM program in October 2013, there were subsequent delays in project preparation and tendering. After a tender process initiated during 2015–2016 failed to elicit bidder interest at the end of the Request for Qualification (RfQ) stage, the process was restarted in 2017. Stringent prequalification criteria led to only four bidders participating and eventually only three bids were received.

Of the three bidders, two bidders were disqualified at the end of the technical evaluation, leaving only one qualified bidder. After an evaluation of the same, the project was awarded to Suez India in November 2017. The PPP agreement was signed between Suez India and Coimbatore Municipal Corporation (CMC), the city government and public counterparty, in January 2018, and the construction commenced in August 2019 after an initial study phase.

Salient features of the public–private partnership arrangement

The scope of work under this contract is spread over 26 years and started with a 1-year study period for design, surveys, and preparation of BOQ, which were done free of charge by the operator. During this period, CMC continued to be responsible for managing the water supply operations. Following the completion of the study period, the operator took over the water supply system and will manage it for a period for 25 years.

The first 4-year period is the construction period. The PPP contract requires the operator to complete construction and system rehabilitation based on the design and BOQ proposed by the operator and approved by CMC during the study period. During the 25-year period, the operator is responsible for O&M of the entire water value chain, starting with the water reservoirs (34 existing and 29 new), distribution network (1,122 km existing and 250 km new), and connections (150,000 at the start). The O&M responsibility also covers meter reading, billing, collection, and customer services.

a. **Revenue model:** The project was bid out based on a composite cost comprising (i) cost of works and (ii) O&M costs. There was no payment made to the operator during the 1-year study period at the start of the contract. The cost of works provided a single estimate for construction, which is to be completed in 4 years. The operator is given flexibility in works, capex phasing, and investment, and in configuring the BOQ, which cannot exceed the cost of works provided in the bid during the study period. These costs are compensated based on the government schedule of rates and/or market rates. The O&M costs (which include return on the operator's investment) are to be quoted as an annual lump sum for each of the 25 years. The revenue covering cost of works and O&M costs is paid as a staggered annuity (in quarterly annuity payments) for each of the 25 years as per an agreed schedule provided by the bidder. Unforeseen asset replacement and major maintenance following completion of the construction period will be compensated separately (see Table 11).

b. **KPIs, incentives, and penalties:** The KPIs under the PPP contract are defined separately for the construction and operation periods. In the construction period, there are two main KPIs linked to operator compensation: (i) installation of customer meters and mapping in GIS and (ii) quality of water treatment at user taps. Both KPIs are linked to penalties in the form of liquidated damages deducted from operator compensation for nonperformance. With respect to the operator's O&M obligations, five performance indicators are monitored and linked with compensation: (i) pressurized supply, (ii) complaint handling, (iii) unaccounted for water (UFW), (iv) water quality, and (v) collection efficiency (see Table 12).

Table 11: Phasing of Annuity Payments in Coimbatore Water Supply Public–Private Partnership

Year	1	2	3	4	5	6	7	8	9	10	11	12	13
Annuity (%)	3.10	4.10	8.50	9.30	1.85	2.00	2.08	2.18	2.28	2.48	2.68	3.00	3.12
Cumulative Annuity (%)	3.10	7.20	15.70	25.00	26.85	28.85	30.93	33.11	35.39	37.87	40.55	43.55	46.67

Year	14	15	16	17	18	19	20	21	22	23	24	25
Annuity (%)	3.21	3.40	3.60	3.70	3.90	4.15	4.65	4.90	5.10	5.30	5.60	5.82
Cumulative Annuity (%)	49.88	53.28	56.88	60.58	64.48	68.63	73.28	78.18	83.28	88.58	94.18	100.00

Source: Copy of PPP Contract December 2017. Coimbatore Water supply. Website of Coimbatore Municipal Corporation.

Table 12: Key Performance Indicators during the Operation and Maintenance Phase

Parameter	Targets	Type of Incentive (I)/Penalties (P)
% of connections converted to continuous 24/7 pressurized supply (applied quarterly)	Min. 7-m water head at ferrule point. Seven days of maintaining continuous supply	I: No incentive P: 1% of monthly O&M in that year for zones not meeting the target
UFW (applied annually)	From 5th to 10th year: 20% From 11th year: 15%	I: One-third of monthly O&M payment for every % point over the target P: One month O&M for every % point less than the target
Resolution of user complaints (applied quarterly)	80% of complaints resolved within a provided time frame.	I: INR 10 million for each % > 90% P: INR 10 million for each % < than 80%
Treated water quality at user taps (applied quarterly)	At WTP: 100% samples as per CPHEEO At consumer tap: 100% samples confirming residual chlorine after COD	I: No incentive P: Rs1,000/- for each nonperformance escalated @ 5% per annum
Collection efficiency (applied annually)	From 5th to 10th year: 80% From 11th year: 90%	I: One-third of monthly O&M payment for every % point over the target P: One month O&M for every % point less than the target

CPHEEO = Central Public Health and Environmental Engineering Organization, O&M = operations and maintenance, UFW = unaccounted for water.

Note: Incentive (I) and Penalty (P) in the table connotes the performance-based incentive and penalty as applicable against the respective parameter and provides the basis for their computation.

Source: PPP Contract Coimbatore Water supply. Coimbatore Municipal Corporation.

Outcomes and the way ahead

Early in the project, some protests about it were reported in the press. However, the construction phase has begun and tangible progress has been achieved. As of October 2020, over 55 km of new pipelines have been laid, and over 10,000 leaks in the network have been identified and corrected, saving an estimated 9 million liters per day (MLD). Construction delayed by the onset of the COVID-19 pandemic should return at full scale soon. A customer service center is also expected to open shortly.

This is among the larger projects carried out in water distribution in recent years in India and follows a period of sharp decline in new project starts during much of the current decade.[15] Although it is premature to conclude, this project along with other smaller PPP projects and DBO contracts awarded could facilitate an increase in momentum for private participation in India's urban water distribution.

Enablers and lessons learned

The project had strong support from the federal and state governments. Capital grants from these governments covered close to 70% of the project cost, which helped address the investment financing challenge. Although a large share of project cost is financed from capital grants, the project's unique staggered annuity compensation format (longer contract period and sharply defined KPIs linked to incentives and penalties) provides sufficient incentive to promote operator accountability and drive efficiency gains.

The strong legal and institutional framework in the state provided enabling conditions. The Tamil Nadu Infrastructure Development Act 2012 (TNID Act), along with attendant rules and regulations, provided a robust legal framework for PPPs and infrastructure development in the state. The Tamil Nadu Infrastructure Development Board (a nodal state-level institution set up under the TNID Act with a mandate to oversee infrastructure PPPs) played a vital role in driving rigor in project preparation and in signaling strong policy commitment to the project.

The project has also benefited from useful project institutional and financial enablers. The water supply function in the city had been fully transferred to CMC. Payment security has been enabled through escrow of a portion of property tax revenues of CMC (earmarked separately as water and sewerage tax) and user charge revenue streams.

Ilkal, India: Performance-based Construct and Operate Contract

The Ilkal Water Supply project was implemented through a performance-based construct and operate contract (PBCOC), a variant of the design-build-operate (DBO) contracting format. Ilkal is one of the 25 cities supported under the ADB-assisted North Karnataka Urban Sector Investment Program. The contracting framework developed with the support of ADB provides a useful template for bringing private sector expertise into design, construction, and operations, particularly where public counterparty capacity to monitor and oversee complex PPP structures is relatively weak. This format has since been used in cities of Karnataka, other Indian states, and countries in South Asia.

---

[15]   In terms of population covered, this is the largest PPP project awarded since the Nagpur city-wide water supply PPP project was awarded in 2011.

Drivers underlying the public–private partnership contract

The project has its origins in the ADB-assisted North Karnataka Urban Sector Investment Program. The project was initiated by ADB with support from the Karnataka Urban Infrastructure Development and Finance Corporation (KUIDFC), the principal agency for urban infrastructure and financing established by the Government of Karnataka. The KUIDFC is the state's partner for urban infrastructure investment programs supported by multilateral agencies and has been involved in the implementation of several other PPP water supply projects in the state.

The project need was driven by the severity of service gaps in Ilkal. Water supply was available for less than 1–2 h, once every 2–3 days during the rainy season, and once every 4–5 days during other seasons. Although 66% of households had a metered connection, water supply was inequitable and customer service was poor. Given the poor services, user charge collections by the municipal council did not cover the costs of supply. Restructuring of water management in the city was required to address these gaps and to modernize and strengthen water services management.

The project was awarded through a two-stage global competitive bid procedure. Eight firms submitted prequalification documents, of which two were prequalified to participate in the bidding. Veolia India was awarded the contract in October 2012 as per the lowest project cost criteria. Construction was completed in 2015, and the operations were completed in 2019.

Salient features of the public–private partnership arrangement

Ilkal municipal council was responsible for providing the detailed design of the project by appointing an engineering consultant. The KUIDFC was responsible for investments and financial planning. The scope of the operator was (i) to carry out the construction works according to the detailed engineering design provided by municipal council and (ii) to operate and maintain the water supply system for 4 years after completion of the construction. Phase I, the construction phase, was spread over 18 months and covered network and system construction and metering at both bulk and customer levels. The preparatory period involved an O&M handover between the city council and the private operator. This was followed by a 48-month O&M period, covering billing monitoring, training employees and Ilkal City Municipal Council staff, management, customer service center setup, connections management, and maintenance.[16]

Under a PBCOC, the KUIDFC was responsible for capital investments. The payments during the construction phase were structured as fixed remuneration based on the BOQ, with a bonus for early completion. During the operations phase, the operator managed all O&M responsibilities and costs, except power costs and staff costs. The operator was paid an O&M compensation, structured in a 60:40 ratio for fixed and performance-based remuneration. KPIs included maintaining a continuous supply to 98% of authorized connections, customer service targets, adequate pressure, and water loss reduction targets. Performance linkages to compensation covered both bonuses and penalties.

The project also had an engineering supervision consultant and an independent technical auditor to support monitoring of the operator's performance. A nongovernment organization carried out education, communication, and information activities for the project. To summarize, the responsibility for capex for the brownfield project was retained by the public sector; however, in line with global trends, a private operator was engaged not only for

---

16   K. Tamaki. 2017. *24/7 Normalized Water Supply through Innovative Public–Private Partnership: Case Study from Ilkal Town, Karnataka, India.* Manila: ADB. https://www.adb.org/sites/default/files/publication/372081/normalized-water-supply-ppp.pdf.

construction but also as a specialist through a performance-based contracting structure. Performance incentives and penalties were introduced to increase accountability and balance risks.

Outcomes and the way ahead

The project helped Ilkal move from an irregular, intermittent supply to a modern, metered continuous pressurized 24/7 supply with performance metrics that compared with best-in-class urban water supply systems. Residential service connections tripled from 3,000 to 10,000, and the user charge collection efficiency increased to over 90% (see Table 13 for project outcomes).

Table 13: Performance Improvement under Ilkal Water Supply

Parameter	2013	2019
Population served	51,000	85,000
Coverage area (%)	57	100
Metered connection (%)	66	100
Electricity consumption	Reduced by about $3–$4 per month at each household.	
Hours of water supplied	Intermittent 1–2 h in 2–3 days during the rainy season and once in 4–5 days during other seasons	24 h
Average pressure in the distribution system (m)	0–1.5 m	14 m
Physical losses (%)	50	8
Response time for customer complaints	NA	Within 24 h
Collection efficiency	NA	> 90%
O&M cost recovery	NA	89%
Number of connections	< 3,000	> 10,000 (Feb 2017)

Source: K. Tamaki. 2017. 24/7 Normalized Water Supply through Innovative Public–Private Partnership: Case Study from Ilkal Town, Karnataka, India. Manila: ADB. https://www.adb.org/sites/default/files/publication/372081/normalized-water-supply-ppp.pdf.

The PBCOC model presents an entry point for small- and mid-tier cities that often find it challenging from an institutional capacity and scale perspective to attract private operators under more complex PPP structures. It can help modernize and strengthen water service management by establishing long-term management practices to put public utilities and municipal departments on a path toward greater cost recovery (footnote 16).

The PBCOC model has been replicated in 12 other cities, and the Government of Karnataka is exploring wider use of it. This model has also been adopted in several other cities in ADB-assisted projects, including in Gaya (Bihar), Tonk (Rajasthan), Pali (Rajasthan), Cossipore (Kolkata), and Dhaka (Bangladesh) with various modifications (footnote 16).

Key enablers and lessons learned

The involvement of ADB in project bid documentation and due diligence played an important role in the project's success. The presence of the state-level nodal agency KUIDFC and its involvement in project preparation, oversight, and monitoring helped to address capacity gaps in the city council.

The PBCOC model is not a conventional PPP but worked well by enabling end-to-end accountability for construction and O&M in a single contractor and building incentives to deliver on both through a sizable O&M period. While the project in Ilkal, a small town, could be delivered with a single contract, larger cities may require multiple contracts considering the scale and complexity of the water network. Strong stakeholder consultations at the start of the project also helped create a positive view of it and wider enthusiasm for it.

The BOQ-based remuneration during the construction phase reduced risks to the operator, which aided in attracting bids from credible private players with operations experience. Greater risk sharing would have dissuaded such operators, given the public counterparty's credit weaknesses. At the same time, remuneration based on performance ensured continuous and dependable service delivery to the customers. Although this was a small-town project, technologies, including smart water technologies, spot billing systems connected to GSM networks, and a Supervisory Control and Data Acquisition system, were deployed effectively.

Jakarta, Indonesia: City-Wide Concessions

Jakarta, along with Manila, was one of the first two cities in Asia to award two PPP concessions: one for its western zone and one for its eastern zone in 1997. After a challenging history impacted by the East Asian financial crisis, the sale of stakes by the original private operators, renegotiations, and public protests and litigation, the projects were pronounced illegal by the country's highest court. The difficulties with Jakarta's water PPPs illustrate the challenge of preparing, awarding, and managing city-level investment-led water concessions.

Drivers underlying the public–private partnership contract

Prior to the award of concessions, PAM Jaya, Jakarta's public water utility, provided access to water supply to only 40% of a served population of about 4 million; its water losses exceed 50%. Private wells were the main source of supply for most of the population. In this context, the Government of Indonesia, in June 1997, awarded two 25-year water concessions. The concession for western Jakarta, covering approximately 2.15 million consumers, was awarded to a subsidiary of Lyonnaise des Eaux (PT PAM Lyonnaise Jaya or Palyja). The concession for eastern Jakarta, covering 2 million consumers, went to a subsidiary of Thames Water International (Thames PAM Jaya or TPJ). These concessions were awarded via a direct negotiated route without any competitive bidding.

Salient aspects of the public–private partnership arrangement

The PPP concessions faced severe challenges from the start:

a. The revenue model for concessionaires was based on an initial water tariff (Rp per kl) along with a formula for its escalation. These terms were arrived at through negotiations rather than a competitive process. The water charge was to be adjusted on a semiannual basis to reflect currency and inflation fluctuations. The concessionaires also assumed the dollar denominated debt of PAM Jaya but did not take on the risk of currency fluctuation. As the East Asian financial crisis unfolded, revenue under the contract declined sharply in dollar terms (from $120 million in 1997 to $22 million in 1998) and made the contracts unviable.

PAM Jaya had neither the resources to compensate for this shortfall nor the ability to increase domestic tariffs to make up for the gap.

b. In the aftermath of the triple challenges of the financial crisis, employee protests, and political upheavals in the early years, the concessions were renegotiated in an agreement that resulted in the creation of the Jakarta Water Supply Regulatory Body (JWSRB). The JWSRB was constituted by the Governor of Jakarta and the Ministry of Public Works as an independent regulator with the authority to review tariff increases and supervise concessions. This helped create some separation between the regulatory and asset holding functions that were hitherto handled by PAM Jaya.

c. A 35% tariff increase was approved in 2001. Following protracted negotiations, water tariffs were brought under an automatic adjustment every 6 months to allow for payment of arrears to the operators. While the automatic tariff adjustment mechanism and rebasing exercises were intended to make the project financially feasible, they were implemented only from 2004 to 2007. In the face of user resistance and opposition, increases in tariffs were stopped in 2007, with the government citing nonachievement of contractual targets by the operators. Both international partners exited the project—Thames Water in 2006 and Suez in 2012—by selling their stakes to other partners. The clamor to terminate contracts grew; the Coalition of Jakarta Residents Opposing Water Privatization (KMMSAJ) launched a petition to protest the private management of Jakarta's water supply in 2011, and in 2013 filed a similar court petition. A protracted legal battle ensued all the way up to Indonesia's Supreme Court, which, in 2017, ordered the government to end the concessions and return water services to the public utility.

Outcomes and the way ahead

Given the volatile history of the project, there have been contradictory reports of performance improvement under the concessions; however, the consensus is that performance on a range of service parameters was suboptimal on several KPIs.

Water tariffs nearly tripled between 1998 and 2010 from Rp 375 to Rp 1,050 ($0.04 to $0.12) per m^3 in the lowest category and from Rp 5,200 to Rp 14,650 ($0.58 to $1.63) in the highest category and grew much faster than per capita incomes in most categories. Water losses were not sufficiently reduced, with the western and eastern regions reporting water losses of 47% and 52%, respectively, in 2007 against the target of 34%.

Although service coverage improved, inequities in supply remained with only 30% of consumers having access to continuous supply. According to a World Bank study in 2009, close to 50% of the population in the western zone and 33% in the eastern zone were not connected to the water network after almost 10 years. The study cited widespread use of private wells, absence of regulation over the use of groundwater, and reluctance of households to pay a connection fee and periodic user charges as further constraining factors.[17]

Although the Supreme Court declared the concessions illegal, it did not explicitly direct the government to cancel the agreements, so the shift from private concessions to PAM Jaya taking full operational control is expected to be protracted. There is a view that this transition could be programmed along with the expiry of concessions slated for 2023. In 2019, the Governor of Jakarta officially announced plans for the administration to retake control of water management from private firms, which is expected to formally kick-start the process of returning control to municipalities.

---

[17] P. Marin. 2009. *Public Private Partnerships for Urban Water Utilities: A Review of Experiences in Developing Countries.* Washington, DC: PPIAF.

Enablers and lessons learned

The Jakarta concessions provide several lessons for structuring and implementing city-wide PPP projects and, importantly, the challenges in tackling complex concession structures.

a. **Importance of competitive bidding and effective regulation:** Network infrastructure systems like water supply are natural monopolies. Determining a competitive price and providing independent regulatory oversight are critical to serve the public interest, ensure transparent and competitive bidding, monitor performance, and pass on efficiency gains to citizens and customers. In complex city-wide concessions, these capacities become vital.

b. **Importance of deep structural and institutional reforms to make concession structures work:** A tariff-dependent city-wide concession requires significant structural and institutional enablers including a strong policy commitment to tariff rationalization and timely revisions; clear performance linkages; and capable, independent regulatory oversight. Concessions, in particular, call for a deep and sustained commitment to structural reform and institutional capacity improvements.

c. **Risk allocation:** The city-wide concessions from the 1990s also highlight the limits of currency risk transfer. The unravelling of the financial crisis affected the city-wide concessions in both Jakarta and Manila. In the case of Manila, however, this was partly mitigated as one of the concessionaires had limited dollar-dominated debt. Box 1 summarizes the progress and outcomes of the Manila water concessions.

d. **Getting the nuts and bolts right:** Large city-wide concessions need mechanisms to mitigate risks around weak information baselines, effective and orderly management and transfer of employees, and intensive stakeholder management. For instance, recent PPP transactions provide for a study period during which the operator works with the public authority to establish the information baseline or plays a role in confirming the BOQ of construction commitments under the project. Similarly, right-of-refusal options to take over public sector employees and staff rationalization through voluntary retirement programs need to be carefully considered upfront. Sharply defined KPIs and effective mechanisms to record, verify, and link them to performance incentives and penalties without ambiguity are also important.

Box 1 summarizes the **Manila water concessions**, which were awarded at around the same time as Jakarta's.

Box 1: Manila Water Concessions

In 1997, Manila awarded two concessions covering its western and eastern regions. The western concession (Maynilad) covered the larger, more developed area with a population of 7 million. The eastern concession (Manila Water) covered 4 million residents and new neighborhoods. The Asian financial crisis began 1 month after the concessionaires took over. Maynilad, which had taken on most of the foreign currency-denominated debt of the previous public utility, ended up in bankruptcy as the Philippine peso lost half of its value.

The western concession was eventually rebid in 2006. The concessions were awarded following a competitive process along with creation of a strong, independent regulator for oversight and tariff setting. Although full coverage was not met as required in the contract as of 2006, access to piped water in Manila expanded significantly within a decade. In the western concession, the coverage expanded from 67% to 86%, and in the eastern zone from 49% to 94%. Over 4 million people gained piped water access between 1997 and 2006, of which around 50% was the result of low-cost community schemes, mainly in the eastern zone. The eastern concession performed better in terms of NRW reduction. As of 2020, the contracts remained operational, although negotiations for contract renewal beyond 2022 were yet to be concluded as of November 2020.

Source: P. Marin. 2009. Public Private Partnerships for Urban Water Utilities: A Review of Experiences in Developing Countries. Washington, DC: World Bank.

Bulk Water and Wastewater Public–Private Partnerships

Relative to water distribution PPPs, which face greater challenges in terms of political risks and public acceptance, PPPs in bulk water treatment and upstream transmission, and in sewage treatment, present lower risk opportunities for attracting private investment. This is especially the case when the PPPs involve bankable counterparties with committed offtake (for water treated in the case of water treatment) or demand (sewage flows in the case of wastewater treatment). The water PPP market of the PRC, for example, has been dominated by upstream treatment facilities of this nature. Indonesia has also focused its PPP activity in the water sector on upstream water supply projects.

It is important to recognize that while PPPs are relatively easy to structure and garner private investment on the promise of annuity payments (linked to performance), these projects add to the fiscal burden in future if not adequately de-risked. For bulk water projects, it is crucial to reform the downstream distribution system to drive operating efficiency improvements (design of metering areas, NRW reduction, and metering) and to promote financial sustainability through tariff rationalization and revision, volumetric billing, and improved collection efficiency. Similarly, sewage treatment plants need efficiently functioning sewerage networks that connect all households, supported through the introduction of user fees.

Umbulan, Indonesia: Build-Operate-Transfer Concession

This project was structured as a build-operate-transfer (BOT) scheme, with a concession period of 25 years. It includes the transmission pipeline, construction of the production system, and offtake for five regencies and cities. The Umbulan project has been categorized as a National Strategic Project as stated in Presidential Regulation No. 3 of 2016 on the Acceleration of National Strategic Project Implementation.

Drivers underlying the public–private partnership contract

The Umbulan Water Supply System project was launched in the 1980s to meet the growing needs for clean water distribution in the East Java province of Indonesia. The Government of East Java province conceptualized and planned the project. Although several studies and tendering processes were carried out between 1988 and 1999, there were many delays between idea generation and implementation. These experiences, however, shaped the eventual conceptualization of the PPP model, and the project was tendered successfully in 2010.[18]

The prequalification for the project was initiated in 2011, which attracted the interest of several international consortiums. After the completion of prequalification in 2012, the five prequalified consortiums were shortlisted and invited to bid. After prolonged bidding, the project was awarded to a consortium of PT Medco Energi International Tbk and PT Bangun Cipta based on the criteria of the lowest fiscal incentive from the Government of Indonesia in 2016.

---

[18]   See Global Infrastructure Hub. https://gihub-webtools.s3.amazonaws.com/umbraco/media/2344/gih_project-preparation_full-document_final_art_web.pdf.

Salient features of the public–private partnership arrangement

Structured as a BOT scheme, with a 25-year concession period, the project involves the construction of the transmission pipeline and production system, and offtake for five cities and regencies (see Figure 3 for the project structure). The project is estimated to cost Rp 2.05 trillion ($140.7 million), 60% of which would be financed by the Government of East Java.[19] The revenue stream for the operator would be sourced from the bulk water payments of the provincially owned water enterprises (PDABs). The PDABs would receive payment from five regional water utilities (PDAMs) that would, in turn, receive the revenue from the water tariffs. Fiscal support for the project includes (i) a viability gap grant of Rp 818 billion ($57 million) from the Ministry of Finance's Viability Gap Fund (VGF) and (ii) a guarantee from the Indonesia Infrastructure Guarantee Fund (IIGF) to mitigate against any failure of a public authority to meet its obligations.[20]

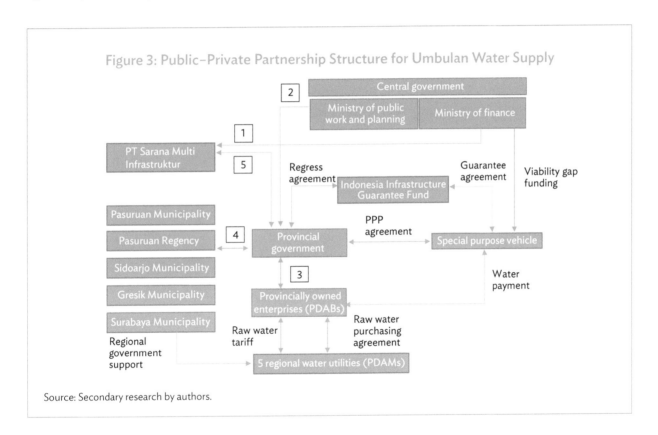

Figure 3: Public–Private Partnership Structure for Umbulan Water Supply

Source: Secondary research by authors.

The operator is expected to deliver 4,000 l of clean water per second through a 93-km transmission pipeline to PDABs, which would distribute the water to five PDAMs responsible for water distribution to customers.[21] The five regional water utilities include Pasuruan Regency, Pasuruan Municipality, Sidoarjo Regency, Surabaya Municipality, and Gresik Regency.

19   Ministry of National Development Planning. 2019. *Public–Private Partnerships- Infrastructure Projects Plan in Indonesia.* https://library.pppknowledgelab.org/documents/5826/download.

20   Global Infrastructure Hub. n.d. *Governmental Processes Facilitating Project Preparation-Indonesia.* https://gihub-webtools.s3.amazonaws.com/umbraco/media/2311/gih_procurement-report_case-study_indonesia_v3_final.pdf.

21   Investment in Indonesia. 2018. *PPP Project Digest.* May. http://investinindonesia.uk/wp-content/uploads/2018/05/PPP-Project-Teaser-roadshow-format-updated-02032018_rev.pdf.

Outcomes and the way ahead

Per press reports, construction was 98% complete, and the project was expected to commence operations shortly. When completed, the project will service approximately 1.3 million people in the designated five districts and cities.

The Umbulan project provides a template for water sector PPPs. Following its award, several other PPPs in the water sector, including the west Semarang water supply project, have been initiated. Indonesia appears to have taken the approach of implementing PPPs in the bulk upstream treatment and transmission segment with offtake commitments from the public water utilities that manage the customer end and downstream distribution.

Key enablers and lessons learned

The project development and implementation have been facilitated by PPP frameworks and regulation, including the Presidential Regulation in 2015, three Ministry of Finance regulations, and two Ministry of Finance decrees. The designation of this project as a National Strategic Project also helped accelerate its development. Its preparation was supported by the Ministry of Finance's Project Development Fund and guided by PT SMI. Viability gap financing from the Ministry of Finance and the guarantee from IIGF also helped in making the project bankable. Another critical aspect was the government's receptivity to feedback. When the project received lukewarm bidder support in earlier rounds, it was rebid with VGF support, and this helped improve the bidding response. Although the bulk water project has been implemented with credit enhancement and support from the provincial government, this needs to be supplemented with reforms at local level, including in distribution, tariff rationalization, and cost recovery.

Sewage Treatment Plants in Cities on the River Ganges, India: Hybrid Annuity

The Ganges is a lifeline for over 500 million people, but over 75% of sewage generated in cities along the river flows untreated into it. Several sewage treatment plants have been developed using a hybrid annuity mode with support from federal and state governments to address the vital treatment infrastructure gaps along the river.

Drivers underlying the public–private partnership contract

The Ganges is 2,525-km long, and serves 43% of India's population. To tackle the challenge of untreated sewage, the Government of India in 2015 approved the "Namami Gange" program. The National Mission for Clean Ganga (NMCG), managed by the Ministry of Water Resources, River Development and Ganga Rejuvenation, developed sewage treatment plants (STPs) through PPPs to treat the pollutants flowing into the sacred river. The International Finance Corporation (IFC) provided transaction support to NMCG, Uttar Pradesh Jal Nigam, and Uttarakhand Pey Jal Nigam for STPs in Varanasi, Mathura (Uttar Pradesh), and Haridwar (Uttarakhand).[22] A single-stage bidding process was conducted to identify the private partner operator, and the lowest bid price was the bid variable in all the projects. Bidding attracted more than 35 companies that participated in the pre-bid consultations. Six bids were submitted for the STP in Haridwar, while Varanasi and Mathura received eight and five bids, respectively. The NMCG, state water authorities, and winning bidders signed agreements in October 2017 to establish STPs under India's first hybrid annuity PPP model for wastewater treatment.

---

[22]   IFC. 2018. *Public–Private Partnership Stories*. Clean Ganges. https://www.ifc.org/wps/wcm/connect/ed4d5a55-c5a3-4acd-b6f1-f26ae80429a3/PPPStories-India-Clean-Ganga.pdf?MOD=AJPERES&CVID=mzDvD.g.

Salient features of the public–private partnership arrangement

The winning bidders have entered a 15-year hybrid annuity contract with the respective state water utility, namely, Uttarakhand Pey Jal Nigam (Planning) and Uttar Pradesh Jal Nigam (Planning). Under this contract, the private operator is responsible for design, construction, commissioning (within 2 years), and O&M of the STPs post-commissioning. According to the IFC, the assets will be transferred to the respective water utility upon completion of the contract term. Under the hybrid annuity structure, 40% of capex is paid to the operator when the construction is completed, while the remaining 60% is paid as annuities over the lifespan of the project along with O&M expenses. The annuity and O&M payments will be conditional on STP performance. The NMCG is responsible for all payments. The state water authority is responsible for monitoring work by the operator during both construction and O&M.

Outcomes and the way ahead

The projects are in the construction phase. KPIs for the operator include (i) availability of the facilities and the associated infrastructure every day during the O&M period, (ii) the quality of treated effluent, and (iii) the sludge consistency. Under the contract, operators can generate revenue in innovative and environmentally sound ways, such as by selling treated wastewater or generating power through biogas. In Mathura, the Indian Oil Corporation will use treated wastewater from the STP for its refinery operations, which would save 20 million liters daily. The structure of the transaction has the potential for replication in other Indian states. NMCG is preparing more hybrid annuity projects in 11 cities using the model documents developed as part of these projects, which is a sign of increased investor confidence.[23]

Key enablers and lessons learned

The hybrid annuity model provides a de-risked way to bring private investment into critical infrastructure requirements.

- **Apex-level stewardship and financing support programmatic replication:** The apex-level coordination and financing support made possible under the NMCG program helped ensure success of the project. Financing from the World Bank through a $1 billion loan also helped secure resources for the development of network and other allied infrastructures. Financing support under the NMCG program helped to make the project bankable and improved its risk profile.
- **Transaction support helped ensure good quality project preparation and a competitive response:** IFC provided technical assistance to NMCG and the state water utilities, which helped with high-quality project preparation, bidding documentation, and bidder interaction support.
- **Concomitant development of network, connectivity, and systemic reforms is critical:** Long-term sustainability will require reforms in policy, institutional capacity, and financial strengthening of the counterparty institutions in the states and local communities. Prompt implementation of sewerage networks and providing incentives for sewage generators to connect to the network will be crucial to ensure that the assets are effectively utilized.

---

[23] IFC. 2018. *Public–Private Partnership Stories*. Clean Ganges. https://www.ifc.org/wps/wcm/connect/ed4d5a55-c5a3-4acd-b6f1-f26ae80429a3/PPPStories-India-Clean-Ganga.pdf?MOD=AJPERES&CVID=mzDvD.g.

Box 2 summarizes key insights from the case studies.

Box 2: Summing Up Key Insights

Water treatment and bulk transmission projects: Indonesia has implemented upstream public–private partnerships (PPPs), including the Umbulan Water Supply project, and has projects in the pipeline in Sembarang, Jatilahur, and elsewhere. These projects are structured using government support mechanisms, including a guarantee from the Indonesia Infrastructure Guarantee Fund, Viability Gap Fund financing from the Ministry of Finance, and offtake commitments. This helps offset tariff risks, although concerns about the pace of distribution reforms still need to be addressed.

Wastewater treatment: PPPs are becoming more common in wastewater treatment. India, for instance, has managed to implement sewage treatment projects in cities along the river Ganges using a hybrid annuity model, where 40% of the project cost is paid upfront and the remaining 60% is paid over the life of the contract and linked to performance outcomes. While it is critical to expedite network implementation to feed these new plants and to achieve utility-level financial reforms, these projects have crowded-in private financing in an under-invested sector.

Water distribution: This segment has seen a variety of contract types, including staggered annuity (Coimbatore, India), concession (Nagpur, India and Manila in the Philippines), and management or lease contracts (Armenia). ADB has supported DBO contracts and performance-based construct and operate contracts in South Asia. Most projects are focused on securing private expertise for operational efficiency and service delivery rather than investment financing. Even where investment financing is prevalent, risks relating to water demand and tariffs are largely off the table in many of the newer PPP contracts in this subsector in the last decade.

Source: Authors.

IV. Three Pivots to Scale Water Public–Private Partnerships and Their Impact

Some of the PPP projects discussed in this publication reaffirm that well-structured and well-implemented PPPs contribute positively to development outcomes. Two aspects become strikingly clear. First, PPPs require an overarching set of enablers to function well and deliver on development outcomes. Second, a "one size fits all" approach will not work. Asian cities are at diverse and unique starting points. They need to contextualize PPP models to fit their specific purpose and policy or institutional ecosystem.

Despite some promising outcomes, mainstreaming water sector PPPs has proven challenging. A wider adoption of PPPs in developing Asia is constrained by three limitations. First, most Asian cities struggle to reconcile water's characteristics as a public good necessity and the economic cost of providing access. Water utilities in Asia rely excessively on fiscal transfers. They have poor cost recovery, often recovering less than their O&M cost obligations. They also have low employee productivity, poor collection efficiency, and high nonrevenue water levels. Second, these policy and institutional weaknesses perpetuate a cycle of low investment–poor service delivery–low cost recovery and weaken public water utilities institutionally and financially. Third, as users resort to coping solutions such as bore wells or tanker supply, informal and unregulated privatization by neglect becomes entrenched and weakens incentives for delivery of public service. Framed against this difficult backdrop, even the few PPPs that are developed become vulnerable to a "set up to fail" scenario.

Changing this status quo will require governments to look at embedding PPPs as an integral element of a wider sector transformation agenda. This will need to be steered along three pivots.

Water Governance Framework

Governance comprises the spectrum of institutions, processes, and procedures that guide government decisions in planning, allocating funds, and implementing public investment projects, including PPPs. The Organisation for Economic Co-operation and Development (OECD) defines the objective of the project governance framework as "… to ensure that infrastructure programs make the right projects happen, in a cost-efficient and affordable manner, that is trusted by users and citizens to take their views into account."[24] There is now a widespread consensus that infrastructure governance is key to achieving value for money in PPP projects. The G20 leaders endorsed the Principles for Quality Infrastructure Investment (QII) at the G20 Finance Ministers' and Central Bank Governors' Meeting in Fukuoka, Japan (8–9 June 2019). The G20 called for stronger infrastructure governance and stressed the importance of assessing value for money and life-cycle costs while ensuring fiscal sustainability.[25]

---

[24] OECD. 2015. *Towards a Framework for the Governance of Infrastructure.*
[25] Available at: https://www.mof.go.jp/english/international_policy/convention/g20/annex6_1.pdf. The G20 agreed on six principles: "(i) maximizing the positive impact of infrastructure to achieve sustainable growth and development; (ii) raising economic efficiency in view of life-cycle costs; (iii) integrating environmental considerations in infrastructure investments;(iv) building resilience against natural disasters and other risks; (v) integrating social considerations in infrastructure investment; and (vi) strengthening infrastructure governance."

The Global Infrastructure Hub published a QII reference guide for PPPs, which includes outcome indicators that water utilities can use to monitor project performance. The OECD has issued guidelines for implementing infrastructure governance frameworks that "guard fiscal sustainability, affordability, and value for money"; ensure a coherent and predictable regulatory framework; and ensure that the asset performs throughout its life in the post-award management phase of the project.[26]

The *Asian Water Development Outlook 2020* (AWDO 2020) observes that most countries have an overarching water policy framework and coordination mechanisms in place. However, the limitation revolves around the following: (i) implementation of water-related policies due to capacity and funding gaps, (ii) insufficient data and monitoring that hampers water policies evaluation, (iii) limited uptake of water policy instruments to manage trade-offs, (iv) limited use of economic instrument to manage water resources, (v) limited effectiveness of regulatory frameworks, and (vi) limited uptake of integrity practices and tools.

The following are the key actions to strengthen the water governance framework:

1.   Renew policy commitment to water security and inclusive access

Although things are changing, policy commitment to drive formal public provision of universal access and service delivery excellence in the water sector has tended to be weak; this is reflected in the large service delivery gaps across water and wastewater sectors. A renewed policy commitment and escalated attention to universal and quality water and wastewater management services is urgent. Critical policy reforms include targeted roadmaps for inclusive and affordable coverage of water and wastewater services, clear tariff and cost recovery policies, and a stable and certain fiscal transfer program (linked to utility performance) for supporting capex and subsidization requirements.

Independent regulation of tariffs and services is critical to fulfilling government's obligations with respect to water as a fundamental public service need. Policies around technical standards for quality in water treatment and supply; wastewater treatment, discharge, and reclamation—both domestic and industrial; equipment, infrastructure, and service delivery are critical as well.

Armenia's PPP program benefited from a strong policy commitment to usher in sector reforms early on. The law on local self-government enacted in 2002 delegated responsibility for providing water supply and sewerage services to local authorities along with the creation of the State Committee for Water Economy (later called the State Committee of Water Systems) and formulation of the water code. The National Water Policy followed in 2005 and the National Water Program in 2006. The Public Services Regulatory Commission and the Yerevan and Armenia water utilities, YWSC and AWSC, were formed. These enabling actions, along with the mobilization of funding support from multilaterals like the World Bank and ADB, were critical steps that brought in private operators to significantly improve water service delivery.

---

[26]   OECD. 2020. *Recommendation of the Council on the Governance of Infrastructure.* Paris. The full set of OECD recommendations are "develop a long-term strategic vision for infrastructure, guard fiscal sustainability, affordability, and value for money, ensure efficient and effective procurement of infrastructure project, transparent, systematic and effective stakeholder participation, coordinate infrastructure policy across levels of government, coherent, predictable, and efficient regulatory framework, a whole government approach to manage threats to integrity, evidence informed decision making, make sure the asset performs throughout its life, strengthen critical infrastructure resilience."

2.   Nurture empowered and capable public counterparties

Water is a local issue; therefore, policy commitment to water needs to be enshrined not only just by federal governments but also by provincial and city governments. Yet across Asia, city governments and urban water utilities are not adequately empowered from either the institutional or the financial perspective. They are thus seen as weak counterparties in PPPs, and this tends to adversely affect PPP project risk profiles. There have been some notable exceptions of strong public utilities in a few cities, including Singapore's Public Utilities Board (PUB), and the Phnom Penh Water Supply Authority, which have delivered strong service delivery performances on their own. Singapore's PUB has used PPPs strategically and successfully in its wastewater reclamation and desalination projects.

Four elements of capacity strengthening among city governments and urban water utilities are particularly critical. Water utilities and local government counterparties need to (i) be empowered with corporate board managed structures and be delegated the powers of independent and autonomous decision-making, (ii) have ring-fenced revenue streams and be well capitalized, (iii) be equipped to attract and retain talent, and (iv) be valued as credible counterparties that can manage PPPs effectively.

Even as governments work to strengthen city-level utilities as credible counterparties, it is helpful to engage state-level nodal institutions as centers of excellence to guide city-level agencies and bridge capacity gaps in project screening preparation, especially in the early stages of PPP programs. However, as the PPP program scales up, improving counterparty capacity is critical.

Clearly, city-level water utilities and local governments that are mandated with water supply need to be adequately empowered with institutional and financial capacities to emerge as credible counterparties for PPP projects. A review of the World Bank PPI database reveals that the counterparties in most water PPPs in Asia are local governments and city authorities, which are often institutionally and financially weak. A clear road map to empower them and equip them with organizational capacity and financial strength is critical for successful PPPs.

Nevertheless, in the early stages of PPP programs, these agencies could benefit from the support of federal and provincial agencies. Funding and project preparation stewardship under the Government of India's NMCG program helped launch sewage treatment plants along the river Ganges in India using a hybrid annuity model. The Coimbatore water supply project received preparation support from the state-level nodal agency, Tamil Nadu Infrastructure Development Board. As an agency empowered under state-level infrastructure legislation, its involvement also had a signaling effect that helped reduce project risk. This form of support could be particularly crucial in small- and mid-tier cities. For instance, the KUIDFC stewarded and supported the PBCOC model of contracting in the Ilkal Water Supply project.

3.   Engender financial vibrancy through revenue reforms and directed subsidies

Financial empowerment is an integral part of institutional strengthening. It is particularly important in the context of PPPs where creditworthiness of the public counterparty is a critical determinant of the project risk profile and bankability. Water utilities across Asia have low cost recovery, often below O&M costs. Implementing a PPP in such circumstances can be particularly challenging, as political and public resistance to tariff rationalization needs to be tackled.

Federal and provincial governments should provide a strong policy direction and support to strengthen policies and implementation arrangements for shifting from flat tariffs and inequitable water access to differentiated volumetric tariffs and equitable supply. Subsidies to deserving segments of the population should be provided in a transparent and directed manner to compensate the utility for the burden.

It is possible to move from levels of poor cost recovery and collection efficiency to sustainable financial performance within a few years, especially if the cycle of low investment-weak service is simultaneously broken. The experience of Ilkal illustrates that when service delivery is improved, collection efficiency and cost recovery improve substantially even in smaller towns.

Enabling Environment

1.   Formulate a sector-level public–private partnership strategy and create a pipeline of bankable projects

PPPs in the water sector are often developed as stand-alone projects and not adequately anchored in a city-province-national water sector strategy. Global experience across an array of sectors suggests that PPPs have flourished in sectors like road building and renewable energy, when developed under the aegis of a sector-level program. This rarely happens in the water sector. Given the scale of the investment needs, PPP development in the water sector must be urgently moved beyond the individual project to evaluate how PPPs can help bridge sector-level deficits.

Federal and provincial governments will need to play a big role in formulating sector-level policy frameworks and PPP strategies. They need to engage cities in a wider dialog and build a pipeline of projects, which can be the focus of rigorous project development support. A strategic thrust in this manner can help inject momentum and scale into PPP programs. The mode of implementation of a PPP project should not be an afterthought but an integral element of sector strategy and project preparation.

A few countries and regions are making efforts on this front. For instance, Indonesia has screened a pipeline of water sector projects to determine their suitability for implementation as PPPs, and some of these projects have moved to the tendering stage. Indonesia's practice of putting together and designating a pipeline of projects as nationally important strategic projects is also a useful approach to build momentum and scale for implementing projects at a sector level. Similarly, the model for sewage treatment plants along the Ganges is being replicated in 11 other cities.

2.   Build rigor in project preparation

Preparation of PPP projects requires rigorous examination often across several stages, including across needs assessment, scope definition, prefeasibility, and detailed feasibility. Furthermore, when projects are to be implemented as PPPs, a comprehensive life-cycle evaluation of costs and benefits through both construction and operations stages is required. Preparation for PPP projects also requires attention to allocation of risk between the public and private party, contract structuring, value for money, and the extent of government support required. As a result, preparation of water PPPs often calls for deep and multifaceted capabilities across technical, financial, environmental, and social aspects.

Water utilities often develop a detailed project design and technical specifications and attempt to overlay and analyze the mode of PPP implementation as an afterthought. This is not an optimal way to develop projects. Countries around the world have evolved methodologies and frameworks for sequencing project development in a structured and systematic manner. Governments can establish clear guidelines and standards for project preparation, project feasibility evaluation, reviews, and approvals to build rigor in project preparation (see Box 3).

Box 3: Project Preparation Stewardship

The **Umbulan Water Supply project** was prepared with support from the Government of Indonesia's Project Development Fund. The Government of Indonesia's public–private partnership (PPP) regulations lay out in some detail the coverage expected from project preparation studies. The project development in this case was managed and overseen by PT SMI, a leading state infrastructure institution.

Similarly, in the case of the **Coimbatore Water Supply project**, the project preparation was carried out following the guidelines and regulations under the TNID Act 2012, which is the provincial government's statute for infrastructure project development in Tamil Nadu, India. Project preparation was overseen by the Tamil Nadu Infrastructure Development Board, the state's nodal agency for project development.

Source: Authors.

3. Build frameworks to provide and manage fiscal support

Almost all the PPP projects discussed in this report required huge amounts of public financing of investment. Even those developed under concessions, like the Umbulan Water Supply project, had sizable upfront viability gap financing. Given the uncertainties in tariff risk and demand risk, often, private operators are reluctant to take on these risks, and therefore, projects are structured using availability payments. The focus of PPPs in the water sector is increasingly on project implementation and efficiency gains, with the private sector contributing a smaller share of investment financing. Under such circumstances, PPPs are no substitute for a large amount of investment; however, they do effectively help governments stagger their obligations by replacing the upfront capital investment with availability payments.

When offered in concert with tariff rationalization and user reforms that yield better cost recovery performance over the medium term, availability payments help reduce a project's risk profile and can be a better solution for governments. However, if availability payments are assured in the absence of reform programs or increased revenues from user fees, the fiscal burden on governments could spiral out of control.

In the last decade, Asian governments have seen their debt increase from 30% to 46% of GDP,[27] and the economic shock caused by the COVID-19 pandemic will make this situation more challenging in the short to medium term. Furthermore, accounting treatments, asset recognition criteria, and weaknesses in contingent liability management frameworks often understate the implications of PPPs on the level of fiscal support needed. For instance, the IMF quantified the average fiscal cost of contingent liabilities for PPPs at 1.2% of GDP (with a maximum of 2% of GDP).[28]

---

[27] World Bank, M.A. Kose et al. 2020, Caught by a Cresting Debt Wave. In *Global Waves of Debt: Causes and Consequences*. Finance & Development, Volume 57. Washington, DC.

[28] IMF, E. Bova et al. 2016. *The Fiscal Costs of Contingent Liabilities: A New Dataset*. Prepared for IMF, January 2016.

Therefore, to effectively manage and support PPP programs of scale, governments will need to back up viability gap grants and availability payments with a comprehensive framework to analyze fiscal costs and contingent liabilities of PPPs (see Box 4).

Box 4: Fiscal Commitments

Public–private partnerships (PPPs) have the potential to offer greater efficiency, private sector finance, and greater value for money when compared with traditional public procurement of infrastructure. However, even under a well-structured PPP, the government bears some risks or provides some financial support, collectively referred to as **fiscal commitments**.

Fiscal commitments may be required to (i) make PPP projects viable and (ii) achieve appropriate risk allocation. The potential advantage of PPP procurement can be erased, and fiscal exposure significantly increased if the fiscal commitments and risks are not thoroughly assessed and managed well. Contingent payment obligations associated with guarantees can also expose the government to fiscal risks that if not appropriately accounted for and managed can put the public debt on an unsustainable path. For these reasons, governments should create a risk management framework to manage their fiscal commitments in a prudent manner.

Fiscal commitments can be either direct or contingent. **Direct liabilities** are payments that are known as regular payments made for availability of the infrastructure service and subsidies that are provided directly from the budget. **Contingent liabilities** refer to payments that are made contingent on the possibility of a future event of which the value and timing are uncertain.

Examples of contingent liabilities include all forms of guarantees given by government, termination payments, and force majeure compensation.

Source: Authors.

Principles for Project Preparation and Management

1. Building a sharp operation and maintenance focus and clear performance linkages to compensation

By its very nature, the focus in a PPP arrangement is to shift the attention and accountability from asset creation solely to service delivery. Therefore, contractual arrangements should ideally have a long O&M period, typically 10 years or more. This is critical in water distribution, especially in brownfield rehabilitation, where results such as NRW reduction require a longer time to materialize consistently. In such projects, it is particularly important to establish a strong baseline on the brownfield network and a shared understanding of the capital expenditure commitments needed. While the Coimbatore case, for example, included a 1-year period for the operator to study and confirm a baseline, other projects like Malviya Nagar and Ilkal provided the BOQs for the capital investment program as part of the bid.

Another critical principle is to ensure that KPIs (i) are verifiable and monitorable, and (ii) can be linked with compensation in the form of incentives, penalties, or both. Linkages to performance orientation are essential for good PPP contracts. A critical aspect is to keep the linkages to a few high-impact KPIs. Having an exceedingly long list of KPIs will make monitoring unwieldy and the linkages prone to contest by both parties.

Many of the cases discussed in this publication illustrate the importance of well-designed KPIs. The first phase of the Yerevan PPP contract had 93 KPIs that were eventually reduced to 25. Even then, in both the cases, only four KPIs were linked to compensation. Similarly, the Coimbatore Water Supply project set out two KPIs during construction and six KPIs for operations that were linked to compensation through incentives and/or penalties.

2. Balancing competition and bidder capability

PPP projects require careful and considered setting of prequalification criteria to maximize competition while ensuring threshold bidder capability. Transparent and fair competition among qualified private developers is essential for efficient price discovery and effective project implementation. The Jakarta Water Supply project, for example, is often criticized for adopting a direct negotiation route for its city-wide concessions, thereby, not bringing completion efficiency and price discovery into the project.

Yet, getting the right balance of competition and capability in water PPPs, especially in the early stages of a country's PPP program, can be difficult; with weak developer-operator ecosystems in the water sector, many Asian countries will need to tap global expertise. If the project is small or the counterparty credit capacity is weak, the response of international bidders tends to be tepid.

Box 5 provides some salient aspects of communication and engagement with bidder ecosystems. The Umbulan Water Supply project is a good example of how project proponents paid attention to the feedback of bidders when there was a lukewarm response to the project without VGF. After factoring in the feedback and incorporating VGF, the project was bid out successfully. Yerevan is another example where the subsequent phase of PPP contracting built on the lessons from the previous stages.

Box 5: Engaging Bidder Interest through an Active Communication Focus

An active communication focus is critical to engage private sector interest. Public–private partnership (PPP) projects are subject to a high level of scrutiny, in general, and during bidding. Communication with bidders will also send a message of transparency and integrity about bidding. Communication actions should facilitate transparent and equitable sharing of information with all stakeholders. To ensure a keen contest, a wide set of potential bidders should be identified early on. It is also critical that the private sector views the project as a viable investment opportunity and the public counterparty as a credible partner that honors its contractual obligations. Early bidder engagement could be through marketing the project to potential bidders and sharing the preliminary information memorandum of the project through road shows and presentations at conferences, and on the website of public counterparty. Pre-bid meetings are a key element of the communication strategy that helps the practitioner build substantial trust and confidence among stakeholders. The pre-bid meetings are an important communication tool of the project implementation agency. A critical aspect of information dissemination is establishment of a data room, where relevant information related to the project is made available equally to all bidders.

Source: Government of India, Department of Economic Affairs. Ministry of Finance. 2016. *PPP Guide for Practitioners.* https://pppinindia. gov.in/documents/20181/33749/PPP+Guide+for+Practitioners/.

3.   Post-award management

The evolution of PPPs in the last couple of decades has provided a strong repository of knowledge and experience to help structure workable PPPs. However, it is critical to build the institutional discipline, integrity, and capacity to operationalize the letter and spirit of the PPP contracts. Post-award management often does not get the attention that it deserves. Two areas should be squarely addressed.

First, the public counterparty should be accountable for administering the PPP contract and ensuring that both parties uphold their obligations. This should be in the form of a designated authority within the public entity who is empowered to make decisions pertaining to the contract and who is supported by a secretariat to provide analytical and monitoring support. Critical aspects, such as payment security, dispute resolution mechanisms, and compensation mechanisms (including linkages with performance obligations), should be well understood and enforced.

Having a provincial-level agency overseeing the contract can be beneficial. For instance, in the Yerevan Water Supply project, the Public Services Regulatory Commission regulates public utilities and their tariffs. In Indonesia, the IIGF guarantees public counterparty obligations. The Tamil Nadu Infrastructure Development Board, the state-level nodal agency, is vested with powers for monitoring PPP projects, and this enables its oversight over the Coimbatore Water Supply project. It is also vested with the ability to approve fiscal support for the project.

Second, given the long contractual periods, PPPs have tended to require renegotiation throughout the tenure of the project. While Latin America has had the most renegotiations[29] over the last couple of decades, these have also occurred frequently in Asia. Between 2005 and 2015, almost one-third of all original contracts were renegotiated. Notwithstanding the moral hazards this imposes on governments, renegotiations often become inevitable in the changing infrastructure landscape. Box 6 captures key principles for renegotiation.

4.   Testing public–private partnership structure for contextual fit and appropriateness

The case studies presented in this article confirm a key principle for good transaction design: the proposed PPP structure needs to be anchored in the project context and needs to be tested for contextual appropriateness. Even in projects that are developed as part of a programmatic initiative, like the STPs along the Ganges, factoring in local and project-specific considerations is crucial.

Possibly the most critical requirement is to ensure that the PPP model chosen is appropriate. For instance, in terms of enabling factors, the prerequisites for a successful tariff-led investment-oriented concession are very different from those needed for an investment-light management contract. For instance, the success of city-wide investment concessions like those implemented in Jakarta or Manila depends upon a very elaborate, sophisticated, capable, and credible regulatory mechanism and framework. Such a mechanism would be relatively difficult to set up for projects in small towns or cities with a limited paying capacity. Similarly, it would be difficult to attract an international bidder of repute to bid for a small project with a very high degree of risk transfer. For example, if the Ilkal project were structured as an investment-led concession, its weak credit standing would

---

29   Global Infrastructure Hub. 2018, Sydney, Australia. *Managing PPP Contracts After Financial Close: Practical Guidance for Governments Managing PPP Contracts, Informed by Real-Life Project Data.*

immediately make the risk profile of the project extremely adverse. By structuring the project as an investment-light PBCOC, Ilkal was able to attract an international operator to bid and implement the project to achieve best in-class service delivery.

Box 6: Guidance on Renegotiation

Renegotiations of public–private partnerships (PPPs) may be used to avoid fiscal controls and push out costs to future government administrations, and by the private sector to recoup funds after underbidding a contract. While renegotiation should be avoided when possible, the long-term nature of many PPPs makes it likely to occur from time to time for justifiable reasons, such as COVID-19. When renegotiations do occur, it is important that the government adopts an appropriate policy and accounting framework for any additional fiscal costs and changes to contract structure related to renegotiation.

Fiscal: Any additional expenditure because of renegotiation should be accounted for and disclosed in the budget. If added costs are material, they should be subject to an independent review outside of the contracting agency and subject to cost-benefit analysis.

Project Preparation and Procurement: Robust procedures for project preparation requiring value for money analysis, competitive procurement, cost-benefit analysis, technical and financial feasibility studies, and environmental and social impact can mitigate risks of renegotiation.

Expert Advice: Financial and economic complexities of PPP contracts may be beyond the expertise of line ministries or new PPP units without significant operational experience. It is recommended under these circumstances for government to contract independent and expert advice for these negotiations with the private sector.

Be Proactive: It is costly to allow problems with a project to grow. The contracting agency should proactively communicate with the private operators to identify problems as early as possible when there is more time and opportunity to solve issues before they become serious problems.

Source: World Bank, 2021. Enhancing Government Effectiveness and Transparency, Chapter 2, Case Study 5: Managing Public–Private Partnership (PPP) Renegotiation, David Bloomgarden.

Conclusion

Universal access to water supply is critical for improving people's quality of life in Asian cities. Tackling water deficits in an ecologically and financially sustainable manner amid the rapid influx of population presents a daunting challenge. However, the cases explored in this study suggest that when effective water governance, policies, and institutional mechanisms are put in place, it is possible to attract private capital and expertise to accelerate the provision of access to quality water supply services.

The cases presented also illustrate that a "one size fits all" approach to PPPs rarely works. Outcomes from private sector engagement are deeply entwined with the underlying water policy, governance, institutional, and societal contexts to which PPPs need to be tailored. A variety of models are available to choose from, depending on where a city's water ecosystem is starting from. However, experience from projects across Asia and globally in emerging economies, including those discussed in this study, provides us with rich insights on what works and what does not.

What is amply clear is that water sector PPPs cannot be developed in isolation. They can rapidly accelerate service delivery performance when implemented against a backdrop of system-wide reform and transformation. Actions identified in this study under the three pivots—creating a strong water governance framework, fostering an enabling environment for PPPs, and embedding good principles of project preparation and management— could help governments, policy makers, and utilities scale up water PPPs and their development impact.

Further Reading

Asian Development Bank (ADB). 2007. *Proposed Loan Republic of Indonesia: West Jakarta Water Supply Development Project.*

———. 2013. *Loan West Jakarta Water Supply Development Project Indonesia.* Extended Annual Review Report.

———. 2019. *Asian Development Outlook Update: Fostering Growth and Inclusion in Asia's Cities.* Manila.

———. 2020. *Karnataka Integrated Urban Water Management Investment Program (Tranche 2).* Semi-Annual Social Monitoring Report.

———. *Karnataka Integrated Urban Water Management Investment Program (Tranche 2).* https://www.adb.org/projects/43253-027/main#project-overview.

ADB Institute. 2019. *Water Insecurity and Sanitation in Asia-Encouraging Private Financing for the Supply of Water through Spillover Tax Revenues.*

Asia-Pacific Economic Cooperation. 2014. *Infrastructure Public–Private Partnership Case Studies of APEC Member Economies.*

Coimbatore City Municipal Corporation. Implementation of 24x7 Water Supply System for the City of Coimbatore. Concession Agreement. https://www.ccmc.gov.pdf.

S. Dasgupta. 2019. 24x7 Water Supply Scheme – The Ilkal Story. Presentation. https://events.development.asia/system/files/materials/2019/03/201903-24x7-water-supply-scheme-ilkal-story.pdf.

Deccan Herald. 2013. *French Firm Bags Water Supply Outsourcing Contract.* 26 June. https://www.deccanherald.com/content/341187/french-firm-bags-water-supply.html.

Fichtner Consulting Engineers India Private Limited. 2017. *Improving & Revamping the Existing / Proposed Water Supply Distribution System with Continuous Pressurized Supply to Coimbatore Corporation.* https://www.ccmc.gov.pdf.

Government of India, Ministry of Water Resources, River Development and Ganga Rejuvenation. 2018. National Mission for Clean Ganga. 11 June. http://smcg-up.org/wp-content/uploads/2019/01/34.-AAES-Mathura-Revised-1.pdf.

Government of Indonesia, Ministry of National Development Planning. 2018. *Public–Private Partnerships-Infrastructure Projects Plan in Indonesia.* https://www.bappenas.go.pdf.

International Finance Corporation (IFC). 2018. *Public–Private Partnership Stories: Clean Ganges (Varanasi, Haridwar & Mathura)*.

IFC. 2017. IFC Helps Structure India's First Hybrid Annuity PPP Project for Sewage Treatment. News release. 11 October.

Japan International Cooperation Agency. 2009. *Preparatory Survey for Public–Private Partnership Infrastructure Project in the Republic of Indonesia.* https://openjicareport.jica.go.jp/pdf/11958063_03.pdf.

Loughborough University. 2003. M. Sohail, ed. *Public Private Partnerships and the Poor-Drinking Water Concessions: A Study for Better Understanding Public–Private Partnerships and Water Provision in Low-Income Settlements.* Jakarta.

Market Research. 2017. SMI—*Umbulan Springs Drinking Water Supply System—East Java—Project Profile.* August. https://www.marketresearch.com/Timetric-v3917/SMI-Umbulan-Springs-Drinking-Water-11100821/.

R. Nugroho. n.d. *Behind the Failed of Jakarta Water Privatization.*

Suez India. 2018. Contracts Scheme for Non-Revenue Water. Asia Water Forum 2018 Presentation. https://events.development.asia/system/files/materials/2018/10/201810-contracts-scheme-non-revenue-water.pdf.

Suez India. n.d. Malviya Nagar Water Services: Improved Services for All. https://www.suez.in/en-in/our-offering/success-stories/our-references/malviya-nagar-water-services-improved-services-for-all.

UN Water. Water and Urbanization. https://www.unwater.org/water-facts/urbanization/ (accessed 6 December 2021).

Veolia India. n.d. Ilkal, Karnataka. https://www.veolia.in/ilkal-karnataka.

World Bank. 2018. *Case Studies: Umbulan Water Supply System Project in East Java, Indonesia.*

———. 2017. *Review of Armenia's Experience with Water Public–Private Partnerships.*

———. 2015. *Evaluation of Water Services Public Private Partnership Options for Mid-sized Cities in India.*

Lightning Source UK Ltd.
Milton Keynes UK
UKHW051239100622
404218UK00012B/126